SpringerBriefs in Earth System Sciences

South America and the Southern Hemisphere

Series editors

Gerrit Lohmann, Universität Bremen, Bremen, Germany

Lawrence A. Mysak, Department of Atmospheric and Oceanic Sciences, McGill University, Montreal, QC, Canada

Justus Notholt, Institute of Environmental Physics, University of Bremen, Bremen, Germany

Jorge Rabassa, Laboratorio de Geomorfología y Cuaternario, CADIC-CONICET, Ushuaia, Tierra del Fuego, Argentina

Vikram Unnithan, Department of Earth and Space Sciences, Jacobs University Bremen, Bremen, Germany

W0225722

SpringerBriefs in Earth System Sciences present concise summaries of cutting-edge research and practical applications. The series focuses on interdisciplinary research linking the lithosphere, atmosphere, biosphere, cryosphere, and hydrosphere building the system earth. It publishes peer-reviewed monographs under the editorial supervision of an international advisory board with the aim to publish 8 to 12 weeks after acceptance. Featuring compact volumes of 50 to 125 pages (approx. 20,000–70,000 words), the series covers a range of content from professional to academic such as:

- A timely reports of state-of-the art analytical techniques
- bridges between new research results
- snapshots of hot and/or emerging topics
- literature reviews
- in-depth case studies

Briefs are published as part of Springer's eBook collection, with millions of users worldwide. In addition, Briefs are available for individual print and electronic purchase.

Briefs are characterized by fast, global electronic dissemination, standard publishing contracts, easy-to-use manuscript preparation and formatting guidelines, and expedited production schedules.

Both solicited and unsolicited manuscripts are considered for publication in this series.

More information about this series at http://www.springer.com/series/10032

Rogelio Daniel Acevedo

Geological Records of the Fuegian Andes Deformed Complex Framed in a Patagonian Orogenic Belt Regional Context

Rogelio Daniel Acevedo
Centro Austral de Investigaciones Científicas
(CADIC-CONICET)
Ushuaia, Tierra del Fuego, Argentina

ISSN 2191-589X   ISSN 2191-5903   (electronic)
SpringerBriefs in Earth System Sciences
ISBN 978-3-030-00165-0   ISBN 978-3-030-00166-7   (eBook)
https://doi.org/10.1007/978-3-030-00166-7

Library of Congress Control Number: 2018953299

This Springer imprint is published by the registered company Springer Nature Switzerland AG
The registered company address is: Gewerbestrasse 11, 6330 Cham, Switzerland

Acknowledgements

To the Master, who always goes with me. Likewise, to Prof. Jorge Rabassa who acted as an efficient translator of the original manuscript, improving it in its syntax and content with his accurate advisory.

Contents

Abstract

The Fuegian Andes show simple and complex aspects in the Argentine portion. The simple aspects refer to the general stratigraphy, and the complex ones result from tectonic characteristics. In a stratigraphic point of view, an appraisal of the denomination and valorization of formations of the Jurassic and Early Cretaceous periods, which has become entangled in recent years with new formational denominations, is performed here, so as to reach the simple scheme of the Fuegian–Patagonian continuity under the name of "Fuegian Andes Deformed Complex." As an essay, the formational nomenclature is simplified and the local concept of the basement is re-evaluated. In the structural aspect, it is assumed that the formation of the orogenic Fuegian arc and the folding of the Mesozoic and Early Tertiary layers are connected phenomena. Geologic forces from the west, the northeast and the south, active in the formation of the tectonic arc, have taken place. The lithostatic column pressure was added once the layers were folded and thrusted upwards during the younger generation of the Fuegian Andes. Even the speculation of a great main fold overturned to the north, as an abstraction, is herein considered. Instead, the Magallanes–Fagnano fault is appreciated only as a product of the transcurrent Quaternary movements without associated eruptive processes that contribute to define plate borders.

Chapter 1
Introduction

Any study of nature confronts the dual attraction between simplicity and complexity. It could be said that the more simple a phenomenon or an object under analysis or observation is, then more difficult it is to state its entity, whereas a complex phenomenon has patterns or characteristics by which it is possible to go ahead and to reach approximations or results. In a few words, simplicity does not mean easiness, as complexity alone does not mean difficulties. In other terms, the geology of the Fuegian Andes Deformed Complex constitutes an oxymoron and both also a paradox of the simpler versus the more complex. Even more, something, that is unique, that is a whole entity itself as the maximum in simplicity terms results unapproachable or unknowable, as it becomes the entire system.

To do investigation in the geological sciences involves a permanent comparative process: things can be equal or different, alike or unlike, connected with their origins or independent of them, homologous or not. Something simple makes difficult the comparative process which is the structural root in the enlightenment mechanism. If a starting point is seen as a simple one, the method of making thins more complex will be the natural logical way of study, stating stages, although they could be transitional, to then use them as comparative terms. And if that starting point is already seen as a complex one, then the method to follow could be that of simplification confronting the risks that, of course, it implies. It is a game of interaction between the two procedures, particularly in front of the series of many interpretations, approaches, or findings made by the many different authors involved with the subject.

Such a concept could be taken in mind in the case of the geology of the Fuegian Andes. They add simplicity plus complexity and that point of view is the direction and the target of this work.

The unquestionable agreement in that the Fuegian Andes is a part, in continuity, of the Patagonian Andes Cordillera is a simple starting point for their interpretation and nomenclature. Thus, e.g., from the Lake Fontana (Chubut) to the Isla de los Estados (Staten Island), there are common terms in the geological history. In this case, the local differences will assume the role of comparative standings.

© The Author(s), under exclusive licence to Springer Nature Switzerland AG 2019
R. D. Acevedo, *Geological Records of the Fuegian Andes Deformed Complex Framed in a Patagonian Orogenic Belt Regional Context*, SpringerBriefs in Earth System Sciences, https://doi.org/10.1007/978-3-030-00166-7_1

Fig. 1.1 Sharp folding in Mount Olivia, a postcard of the Ushuaia landscape. *Credit* E. Pocai

This is a simple premise versus the complex matter of the acute folding attached to a low metamorphism degree in the singular environment of the Fuegian orogenic arc.

They are two different studies, each one with their own characteristics concerning the degree of complexity, and both demanding that each one has no impact on the other. That is to say, the structural complexity must not blur the clear acceptance of the geological condition of the Patagonian Cordillera (Folguera et al. 2016 and references therein). But it is understood that the pervasive deformation (Fig. 1.1), which implies repetition of the folded strata, and the metamorphism, although of a low grade, obscures the definition of the formations and the final statement of their own formational chart.

These worries were made visible in the recognition of the Fuegian Andes Deformed Complex (Fig. 1.2), formerly advanced as *"Complejo Deformado de los Andes Fueguinos"* (Quartino et al. 1987; Acevedo 1988) in the idea to join, in one denomination, the two aspects already mentioned without tackling the definition of the formations, a subject which will be enlightened with the progress of more detailed studies, surpassing the objectives of the present investigations.

These works move further away from the risk of the generalization and they become milestones in the way toward the organization of the subject as a whole. Such research has been performed for a long time since, as in the contributions by Camacho (1948), Furque (1966), Borrello (1967, 1969), Caminos et al. (1981), Quartino et al. (1989), Olivero and Martinioni (2001), among others, as we will see later on (González-Guillot et al. 2012), an approach that has continued until the present time with the last updating version by Ghiglione (2016).

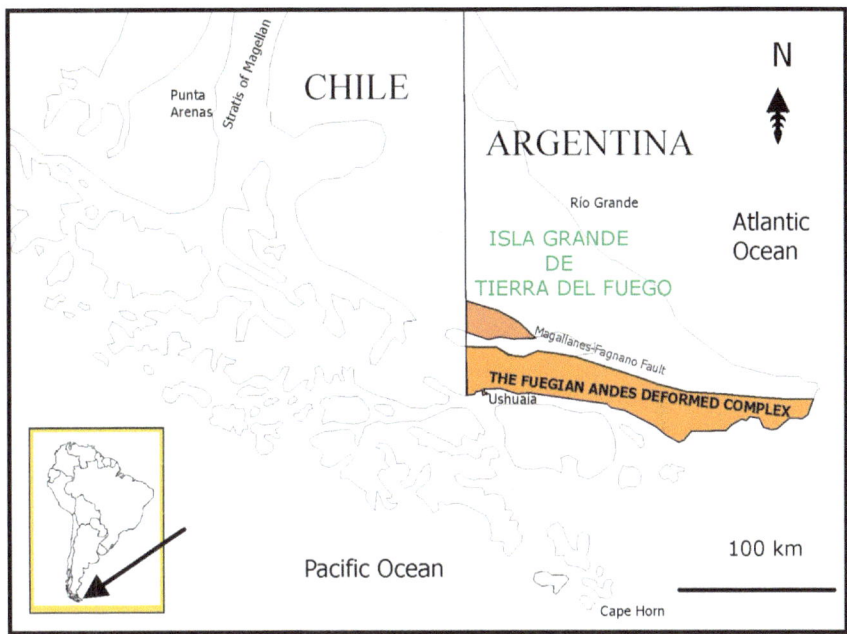

Fig. 1.2 Geographical position of the Fuegian Andes Deformed Complex

In connection with the above statements and having in mind the complexity which is added to the stratigraphic order, in this book there will be the addition of certain considerations about the origin of the Fuegian tectonic arc and the penetrating folding of the associated layers.

Figure 1.3 contains some geographical locations that are mentioned in the text throughout the work.

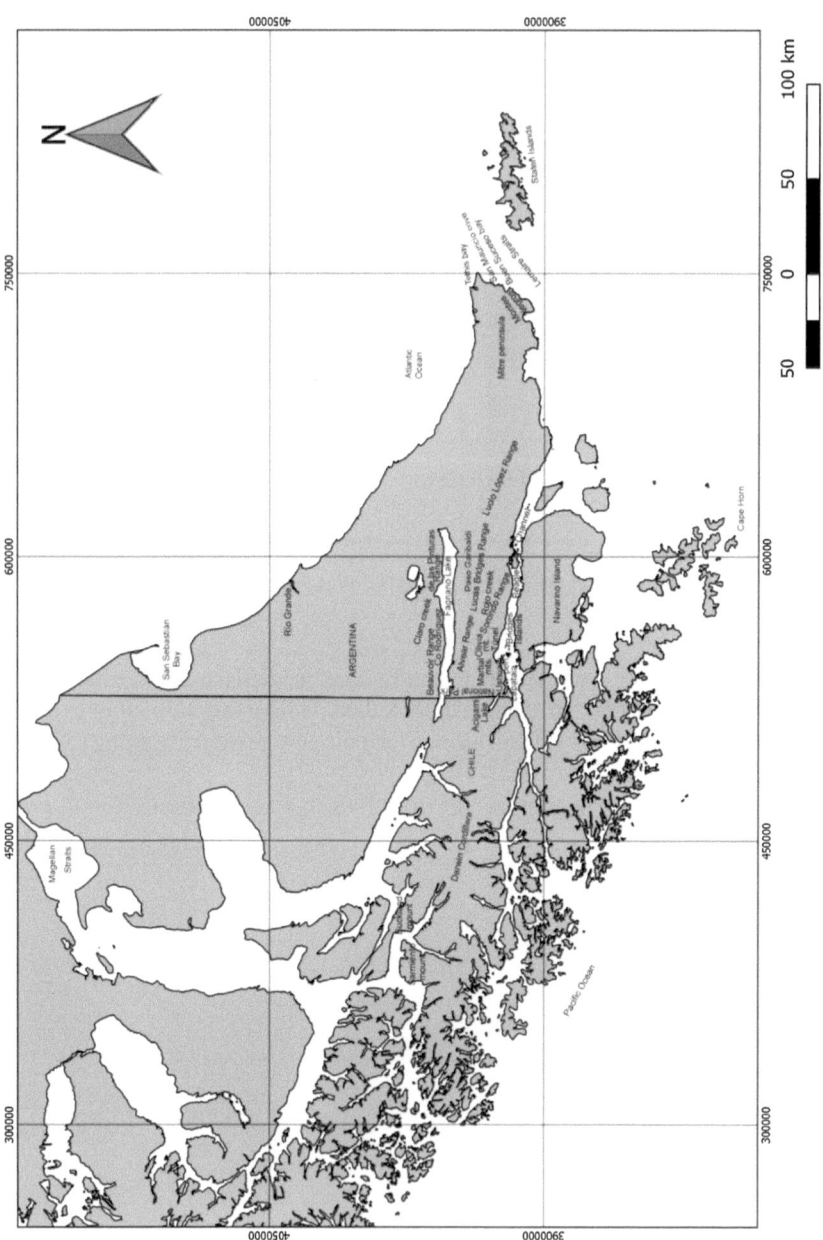

Fig. 1.3 Geographical references mentioned in the text

References

Acevedo RD (1988) Estudios geológicos areales y petroestructurales en el Complejo Deformado de los Andes Fueguinos. Universidad Nacional de Buenos Aires. Biblioteca de la Facultad de Ciencias Exactas y Naturales, Ph. Thesis, 233p

Borrello AV (1967) Estado actual del conocimiento geológico del flysch en la Argentina. Revista del Museo de La Plata. Tomo VI. Geología, Nº 44:125–153 (La Plata)

Borrello AV (1969) Los geosinclinales de la Argentina. Dirección Nacional de Geología y Minería, Anales 14 (Buenos Aires)

Camacho HH (1948) Geología del Lago Fagnano o Cami. Universidad Nacional de Buenos Aires. Biblioteca de la Facultad de Ciencias Exactas y Naturales. Ph. Thesis nº 543

Caminos R, Haller M, Lapido O, Lizuain A, Page R, Ramos V (1981) Reconocimiento geológico de los Andes Fueguinos. Territorio Nacional de Tierra del Fuego. 8º Congreso Geológico Argentino 3:754–786 (San Luis)

Folguera A, Naipauer M, Sagripanti L, Ghiglione M, Orts DL, Giambiagi L (2016) Growth of the Southern Andes. Springer Earth System Sciences, 308p

Furque G (1966) Algunos aspectos de la geología de Bahía Aguirre, Tierra del Fuego. Revista de la Asociación Geológica Argentina 21(1):61–66 (Buenos Aires)

Ghiglione M (2016) Geodynamic Evolution of the Southernmost Andes. Connections with the Scotia Arc. Springer Earth System Sciences, 206p

González-Guillot M, Prezzi C, Acevedo RD, Escayola M (2012) A comparative study of two rear-arc plutons and implications for the Fuegian Andes tectonic evolution: Mount Kranck Pluton and Jeu-Jepén Monzonite, Argentina. J S Am Earth Sci 38:71–88

Olivero EB, Martinioni DR (2001) A review of the geology of the Argentinian Fuegian Andes. J S Am Earth Sci 14:175–188

Quartino BJ, Acevedo RD, Scalabrini Ortiz J (1987) Rocas eruptivas vulcanógenas entre Monte Olivia y Paso Garibaldi, Isla Grande de Tierra del Fuego. Simposio Internacional de Vulcanismo Andino, Actas IV:209–212 (San Miguel del Tucumán)

Quartino BJ, Acevedo RD, Scalabrini Ortiz J (1989) Rocas eruptivas volcanógenas entre Monte Olivia y Paso Garibaldi, Isla Grande de Tierra del Fuego. Revista de la Asociación Geológica Argentina 44(3–4):328–335 (Buenos Aires)

Chapter 2
Background

The study of the geologic features of the Argentinian Fuegian Andes has been rich in results and publications both from local and foreign scientists. Although Charles Darwin had already been here too, the list finds its center in the exploration and research made by Kranck in 1932a (Fig. 2.1; see also Kranck 1930a, b, c, 1932b, c, 1934), and it is so much important that the reference and appreciation of his work are never missing in later contributions, since those mentioned above have provided with important new data and points of view.

Previously, the first stage of research and studies in Argentina comprised the works by Bonarelli (1917) and his simpler but very useful geological map (Fig. 2.2), and the ones by Harrington (1943), concerning his observations in Isla de los Estados, which characterized the deformation of the acid volcanic rocks and built a stratigraphic chart which is still valid today, despite the original geological nomenclature has been replaced by modern formational names.

The investigations made progress under the lead of the *Dirección de Minería y Geología de Buenos Aires,* which was then headed by Petersen (1949), in which the geologists, at that time young scientists that would contribute later with important works like Guillermo Furque and Horacio Camacho were involved and whose contributions are a must to be consulted at present.

There were then several other studies until the 70s and the 80s when two Argentine institutions put their interest in the surficial geology of Tierra del Fuego. They were the former *Dirección de Minas y Geología*, or *Secretaría de Minería*, in the Isla de los Estados (Caminos 1975, Caminos and Nullo 1979) and in the Isla Grande de Tierra del Fuego (Caminos et al. 1981; Fig. 2.3) and the *Consejo Nacional de Investigaciones Científicas y Técnicas* (CONICET), which made its research through the *Centro de Investigaciones en Recursos Geológicos* (CIRGEO), with the help of the *Museo de Ushuaia* and the *Centro Austral de Investigaciones Científicas* (CADIC).

It should be mentioned here as well the contribution of Caminos (1980) related to the description and interpretation of the Fuegian Cordillera in the "*Segundo Simposio de Geología Regional Argentina.*"

© The Author(s), under exclusive licence to Springer Nature Switzerland AG 2019   7
R. D. Acevedo, *Geological Records of the Fuegian Andes Deformed Complex Framed in a Patagonian Orogenic Belt Regional Context*, SpringerBriefs in Earth System Sciences, https://doi.org/10.1007/978-3-030-00166-7_2

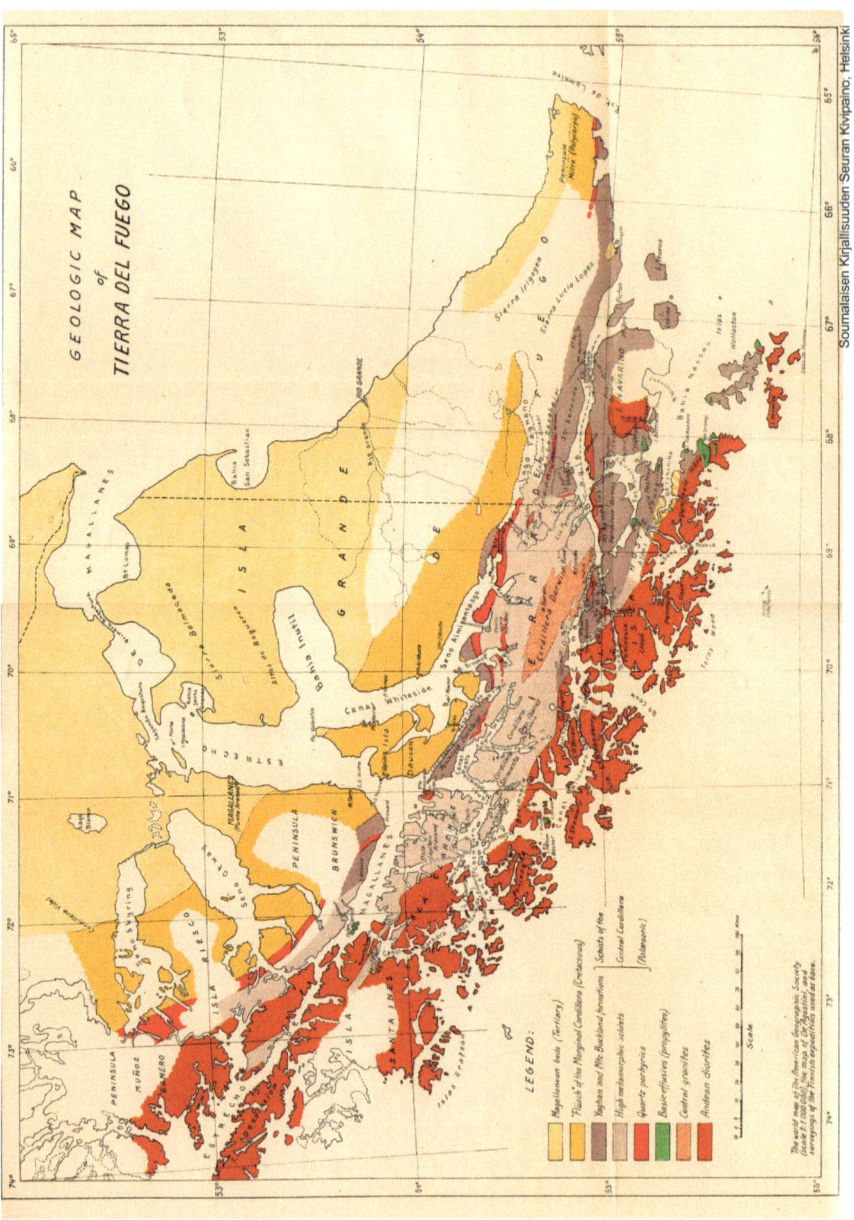

Fig. 2.1 Reproduction of the original geological map of Tierra del Fuego by Kranck (1932a)

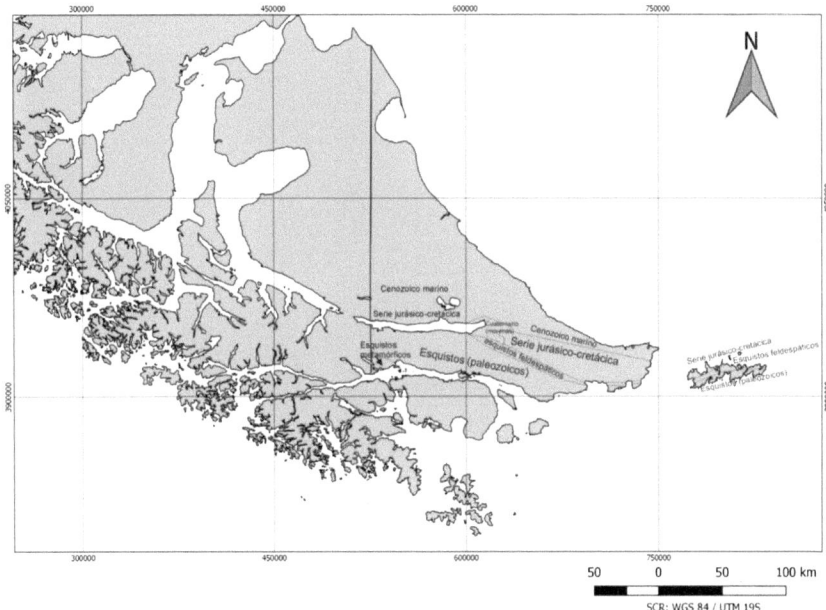

Fig. 2.2 Modified sketch map of the geological outline of Tierra del Fuego by Bonarelli (1917)

The relationship between what was going to be the CIRGEO with the geology of Tierra del Fuego was born in the *Facultad de Ciencias Exactas y Naturales*, University of Buenos Aires, in the 60s when Tomás Suero contacted with B. J. Quartino (verbal communication) concerning some samples from the Alvear Range, at the center portion of the Fuegian Andes. They were highly deformed rocks with mylonitic structure or schists of low grade with sharp isocline folding and contorned *ptigmas* at a detailed scale (Fig. 2.4). Those were verified years later in the Martial Mountains, being this the place where Kranck mentioned them in 1932.

The academic and scientific activity of CADIC-CONICET (Fig. 2.5) in Tierra del Fuego began in the 80s with the development of the thesis of Acevedo (1988) and Bujalesky (1990) and continued during the present century with the doctoral theses by González-Guillot (2009), Martinioni (2010) and Torres Carbonell (2010), and new perspectives are opened with the founding and projection of the University of Tierra del Fuego.

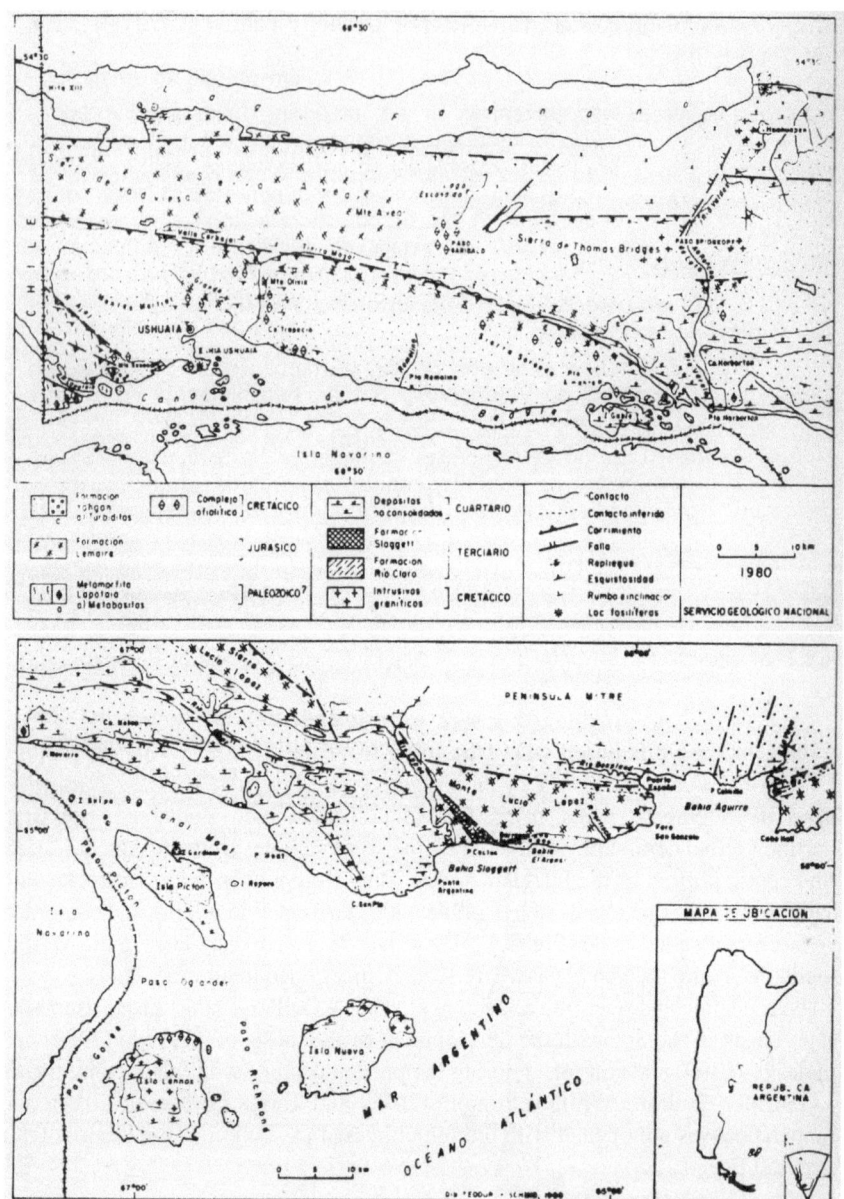

Fig. 2.3 Original geological maps presented by the Secretaría de Minería research team in the VIII Congreso Geológico Argentino (by Caminos et al. 1981)

Fig. 2.4 Isoclinal folds in the Alvear Range ($\lambda = 3$ cm) (left). *Credit* R. D. Acevedo

Fig. 2.5 Aerial view of the Centro Austral de Investigaciones Científicas (yellow buildings at the centre of the image), the southernmost institute of the Consejo Nacional de Investigaciones Científicas (CONICET) in Ushuaia, Tierra del Fuego. *Credit* CADIC-CONICET. It is also the southernmost research center of the World with permanent scientific staff

References

Acevedo RD (1988) Estudios geológicos areales y petroestructurales en el Complejo Deformado de los Andes Fueguinos. Universidad Nacional de Buenos Aires. Biblioteca de la Facultad de Ciencias Exactas y Naturales. Ph.D. Thesis. 233pp

Bonarelli G (1917) Tierra del Fuego y sus turberas. Anales del Ministerio de Agricultura de la Nación. Sección Geología, Mineralogía y Minería, XII, N° 3 (Buenos Aires)

Bujalesky G (1990) Morfología y dinámica de la sedimentación costera en la península El Páramo, bahía San Sebastián, Isla grande de la Tierra del Fuego. Universidad Nacional de La Plata. Facultad de Ciencias Naturales y Museo, Ph.D. Thesis n° 0562

Caminos R (1975) Tobas y pórfiros dinamometamorfizados de la Isla de los Estados, Tierra del Fuego. 6° Congreso Geológico Argentino, 2:9–23 (Bahía Blanca)

Caminos R (1980) Cordillera Fueguina. En Geología Regional Argentina. Academia Nacional de Ciencias. Córdoba, II:1463–1501

Caminos R, Nullo F (1979) Descripción geológica de la Hoja 67e: Isla de los Estados. Servicio Geológico Nacional, Boletín n° 175 (Buenos Aires)

Caminos R, Haller M, Lapido O, Lizuain A, Page R, Ramos V 1981. Reconocimiento geológico de los Andes Fueguinos. Territorio Nacional de Tierra del Fuego. 8° Congreso Geológico Argentino, 3:754–786 (San Luis)

González-Guillot M (2009) Estudio petrogenético de plutones de la cordillera Fueguina, entre el lago Fagnano y el canal Beagle y algunas consideraciones sobre las mineralizaciones asociadas. Universidad Nacional de La Plata, Ph.D. Thesis. 327pp

Harrington HJ (1943) Observaciones geológicas en la Isla de los Estados. Museo Argentino de Ciencias Naturales. Anales, Tomo XLI. Geología. Publicación N° 29:29–52 (Buenos Aires)

Kranck EH (1930a) Sur l age de la Cordillere de Magellan. Extrait du Compte Rendu Sommaire des Séances de la Societe Geologique de France, fascicule 6–7:67-68

Kranck EH (1930b) Sur le profil longitudinal de la Cordillere de la Terre de Feu. Extrait du Compte Rendu Sommaire des Séances de la Societe Geologique de France, fascicule 9–10:102–103

Kranck EH (1930c) Om förekomsten av nyttiga mineral i magallanesländerna. Särtryck ur Geografiska Sällskapets i Finland Tidskrift, 1–2

Kranck EH (1932a) Geological investigations in the Cordillera of Tierra del Fuego. Acta Geogr 4(2):231p (Helsinki)

Kranck EH (1932b) Sur quelques roches a radiolaires de la Terre de Feu. Extrait du Bulletin de la Societe geologique de France, 5ᵉ serie, 2:275–283

Kranck EH (1932c) Sur quelques roches à Radiolaires de la Terre de Feu. Geologisch-paläontologisches Institut der Universität Basel. Societe Geologique de France, 275–283

Kranck EH (1934) The South Antillean Ridge. Bulletin de la Commission Geologique de Finlande, 104:99–103

Martinioni DR (2010) Estratigrafía y sedimentología del Mesozoico Superior-Paleógeno de la Sierra de Beauvoir y adyacencias, Isla Grande de Tierra del Fuego, Argentina. Universidad Nacional de Buenos Aires. Biblioteca de la Facultad de Ciencias Exactas y Naturales. Ph.D. Thesis. 196pp

Petersen CS (1949) Informe sobre los trabajos de relevamiento geológico efectuados en Tierra del Fuego entre 1945 y 1948. Dirección General de Industria y Minería (Buenos Aires)

Torres Carbonell P (2010). Control tectónico en la estratigrafía y sedimentología de secuencias sinorogénicas del Cretácico Superior-Paleógeno de la faja corrida y plegada Fueguina. Universidad Nacional del Sur. Ph.D. Thesis

Chapter 3
Western Sector of the Fuegian Andes Deformed Complex (FADC-WS)

3.1 The Formations in the Mesozoic Complex

The Mesozoic complex starts here from the simplest chronostratigraphic chart, consisting of a thick marine layer of Late Jurassic–Early Cretaceous age, superimposed to Jurassic volcanic and sedimentary rocks and overlain by Late Cretaceous marine sedimentary rocks, the whole sequence folded and metamorphosed. The marine Tertiary, folded as well, is overlying all the aforementioned units. Underlying the whole group, there is a possible metamorphic basement.

In this sense, several formations have been proposed, what in certain ways make much more complex the matter at the same time that they contribute to build a richer system of order and classification.

3.1.1 The Yahgan Formation

As the equivalent of the "Clay Slate Formation" from Gregory (1915), Tyrrell (1930), and Wilckens (1930), this stratigraphic name has priority as being stated in 1932 by E. Kranck. It still enjoys a good bibliographic health, although a newly accepted criterion has modifying its estimated age, initially stated as Paleozoic (Bonarelli 1917, Kranck 1932a), then Cretaceous in the region of Canal Beagle (Beagle Channel) (Richter 1925), now related to Tithonian-early Cretaceous interval. It is worthy to remark here, from a merely cultural point of view, that this is the only Fuegian formation in the territory of Argentina having a Native American name, although it could be much more appropriated the name of "Yamana Formation," which matches the original and correct primitive voice.

The validity of the *Yahgan Fm.* holds both positive aspects and, on the opposite, enigmatic and doubtful ones. The first is related to its age and to the finding of the contact with the underlying rocks. It is the reconnaissance by Kranck in 1932a

© The Author(s), under exclusive licence to Springer Nature Switzerland AG 2019

R. D. Acevedo, *Geological Records of the Fuegian Andes Deformed Complex Framed in a Patagonian Orogenic Belt Regional Context*, SpringerBriefs in Earth System Sciences, https://doi.org/10.1007/978-3-030-00166-7_3

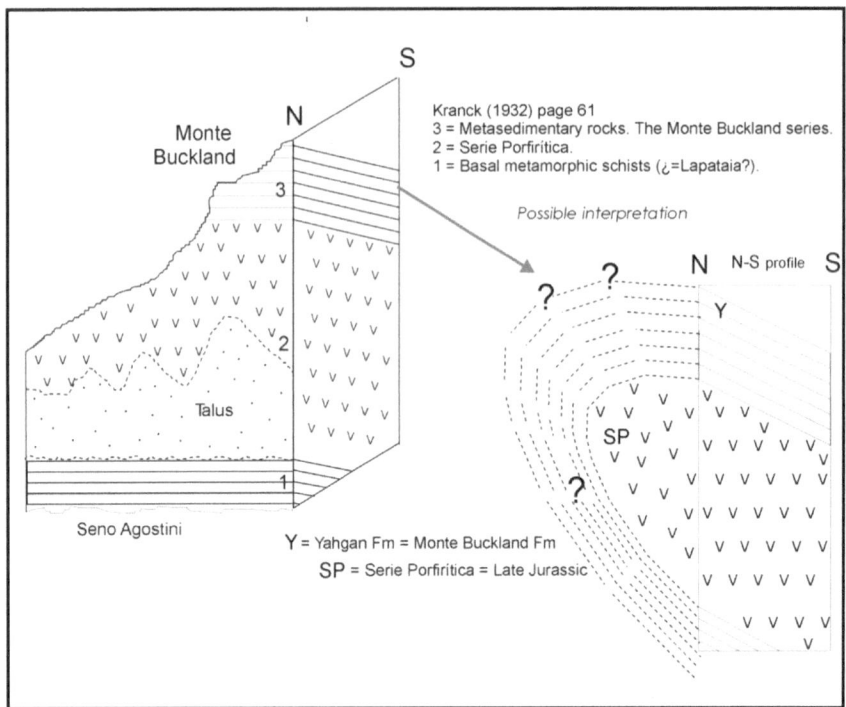

Fig. 3.1 Possible interpretation by the present author of the *Yahgan Fm.* in relation to the description of the Monte Buckland formation made by Kranck (1932a) within Chilean territory

(p. 61) in Monte Buckland, Chile, of clayey metasedimentary rocks or felsic and phyllitic schists located over mylonitized quartz porphyries (Fig. 3.1). Those, or the one known as the "*Monte Buckland series*", would simply be the *Yahgan Formation* and then the Basal Metamorphic Schists (=Lapataia Formation?) would be then, at least in part, a portion of the *Yahgan Formation* as well.

Concerning the existence of reliable chronostratigraphic data, they are very scattered and they come from those areas near Canal Beagle. Thus, the Tithonian–Neocomian units in Chile (Aguirre Urreta and Suárez 1985) and the Aptian–Albian ones in Argentina (Dott et al. 1977, Olivero and Martinioni 1996) should be mentioned.

The unconvincing aspects are: (1) the absence or lack of finding of guide fossils in wide areas of this formation and (2) existing doubts concerning the extension in area of the formation in view that the statements by Kranck in the Argentinian territory are limited to an area close to Ushuaia.

Point one is critical in the type area of the formation which are the surroundings of Ushuaia. Thus, it is easy to understand the cautious criterion used by Borrello (1969) to name as Monte Olivia Formation the metasedimentary rocks of the Olivia and Martial mounts being unsure about a correlation between them. Precisely at Mount Martial, the vertical isocline folding presents a great difficulty for any speculation

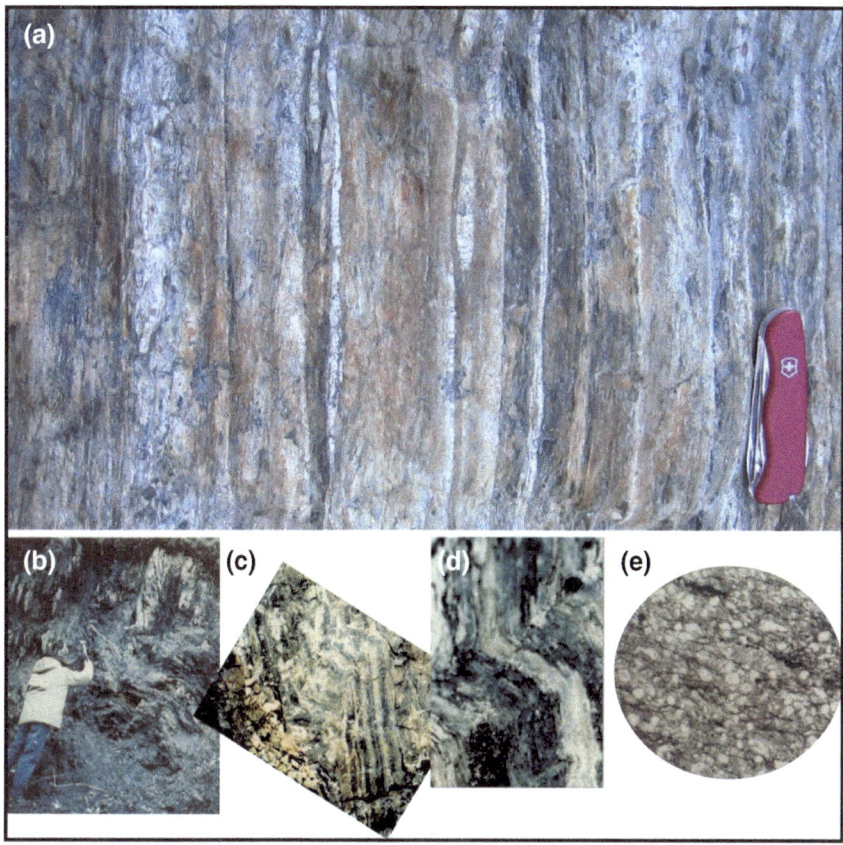

Fig. 3.2 a Typical sequence of pelites and greywackes of the *Yahgan Fm.* in Mount Olivia; monoclinal folds kink bands, macro (**b**), meso (**c, d**), and micro (**e**: ftanite, radiolarite, magnification × 10, PPL) scales from left to right. *Credit* R. D. Acevedo

about the thickness because it is placed in an environment of turnover disharmonic foldings, of many orders, of metapelites and metasandstones, ftanites (Fig. 3.2) and less frequently, impure limestones. Such structure, which dips toward the South, is clearly visible in the oblique or transverse gorges to the local ranges like those visible in the upper course of the Olivia River (Cenni 2005), and the Andorra valley near Ushuaia.

The metamorphism of the *Yahgan Fm.* is of low grade being lower in the greenschist facies, at the zeolite level with the presence of prehnite (Fig. 3.3) and pumpellyite (Caminos 1980), with predominance in the deformation over the recrystalization and neomineralization, which has influence reducing or, at least keeping at the same level, the finer grainsize. Those features as the regional abundance of metamudstones increase the above-mentioned difficulties concerning the follow up of the formations. Added to the aforementioned information is the isolated verifica-

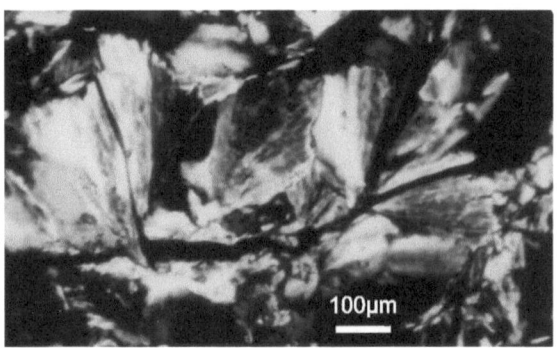

Fig. 3.3 Prehnite–pumpellyite schists of *Yahgan Fm*. XPL. *Credit* R. D. Acevedo

tion of contact metamorphism that has produced neomineralization and a rise in the metapelite compaction together with a decrease in its fissility. This fact has happened, i.e., as consequence of the ultramafic intrusion of *Estancia Túnel* (Túnel Farm) in *Bahía Lancha Packewaia* (Lancha Packewaia Bay), a few kilometers to the East of Ushuaia, close to Canal Beagle (Acevedo et al. 1989, 2002), where there are some hornfels visible in the field next to the contact with the plutonic rocks. Such thermal effect continues far away toward the Olivia River. There, only under the microscope it may be possible to confirm this metamorphism, except for a higher compaction in the rocks. It has produced sericitic recrystallization of pyrite and erratic formation of biotite, garnet, and magnetite.

It is possible to assume that in this area, the plutonic rocks continue below the surface, appreciating the importance of this phenomenon that has influence in the physical appearance of the rocks, particularly because there are many other intrusions and some of them may have not been exposed by erosion. Thus, we have the case of the Ushuaia Hornblendite found at 54° 49′S, 68° 10′W (Acevedo et al. 1989, Acevedo 1990, 1992a, 1996, Elsztein 2004) (Fig. 3.4), also named as the Ushuaia Pluton (González-Guillot et al. 2018); the Jeu-Jepén Diorite (54° 35′S, 67° 13′W, Acevedo et al. 2000, 2004), also known as the Jeu-Jepén Monzonite (González-Guillot et al. 2012), the Puente Quemado Gabbro (54° 50′S, 68° 27′W; Villar et al. 2007, González-Guillot et al. 2016), the Moat Diorite Pluton (54° 46′S, 66° 57′W; González-Guillot et al. 2009) (Fig. 3.5), the Rancho Lata Gabbro (54° 41′S, 67° 15′W; González-Guillot et al. 2010), and the Mount Kranck Pluton (54° 33′S, 68° 09′W; González-Guillot et al. 2012).

It is worth mentioning here that the ultramafic rocks of *Estancia Túnel* were regarded (Caminos et al. 1981) as ophiolites, and at a point of view, it is not shared here. Such discussion is certainly either an opinion or an hypothesis, and as a consequence, it deserves to be respectful, but it must be stressed here that at the time of the search for ophiolites, not in the petrographic sense as Steinmann (1927) or Coleman (1977) but ophiolites as part of the oceanic floor, and as a result of plate tectonics, it was a critical subject of research in the proposal of the hypothesis itself. The above-mentioned appreciation was complemented by the understanding that the

Fig. 3.4 Small plutonic apophysis intruded in metasedimentary rocks in the Ushuaia peninsula. *Credit* C. Elsztein

Fig. 3.5 Natural quarry of intrusive blocks in the Lucio López mountain range. *Credit* M. G. Guillot

basic dikes located north of *Sierra Sorondo* (i.e., Sorondo mountain range) were in fact sheeted dikes of the ophiolitic column.

The second problematic aspect of the *Yahgan Fm.* is the extension of its outcrops, which brings us to the matter of its relationships with other formations.

Fig. 3.6 Convolute bedding of *Yahgan* greywacke levels in the Bridges Islands. *Credit* R. D. Acevedo

Outcrops in the Isla Grande and in some of the smaller islands of Canal Beagle were visited (Acevedo 1988), confirming their recognition as part of the *Yahgan Fm.* (Fig. 3.6). It was decided to leave as future task a more detailed analysis of the rocks which are possibly interdigitated with the *Lemaire Fm.* to the east of Mount Olivia (Biel et al. 2007), in the south margin of the Lashifashaj depression along National Route 3, and the relationships of the upper strata with the proposed *Beauvoir Fm.* to the north face of the Alvear Range.

The aforementioned section of National Route 3 arises the option between the *Yahgan Fm.* and the "*Serie Porfirítica*" or Lemaire Fm. Lithostratigraphic units, a subject which has already been discussed (Biel et al. 2007, and references therein), where it is commonly seen, like at the sources of the Arroyo Encajonado (Encajonado creek) and its tributary Arroyo Rojo (Rojo creek), in the core of the mountains, the common acute folding and overthrusting phenomena of the leptometamorphic schists and volcanic sequence. These rocks comprise the borders of the southern hillsides of the Sorondo Range, all of them located on a thick deformed sequence of rhyolites and basalts located to the north (Iovine 2005).

Both the *Yahgan* Formation and the *Beauvoir* Formation belong to the Early Cretaceous, a fact which arises the matter of their self-identity or their amalgamation as a single entity. Looking for simplicity, it does not seem appropriate to separate the two formations, which are supposed to be of the same age, located in the same area and which are also difficult to characterize. They were kept as independent entities after the statement of the *Beauvoir Formation,* a term coined by Camacho (1948) in the regional environment during his paleontological research at Hito XIX (Fig. 3.7) but it was founded on the identity of a foreign fossil that someone gave him as obtained from the supposed type locality (H. H. Camacho, personal communication).

Fig. 3.7 Modified geological sketch map presented in H. H. Camacho's doctoral thesis (1948)

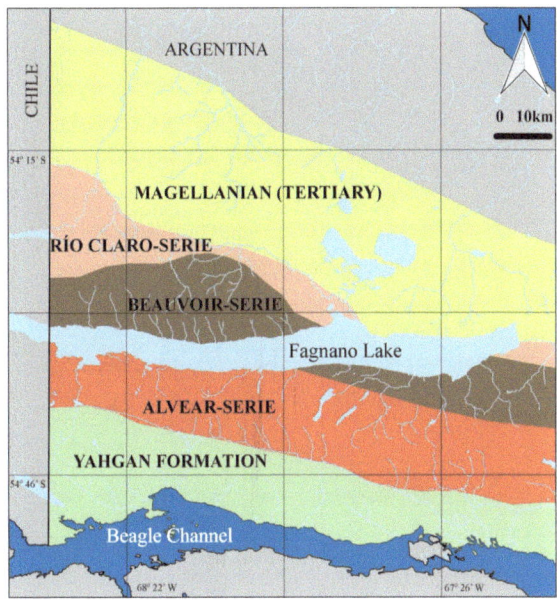

Thus, the concept of *Beauvoir Fm.* persisted until the review work by Olivero et al. (1999) in which the *Yahgan Fm.* is interpreted as the lateral equivalent of the *Beauvoir Fm.–Hito XIX Fm.,* conveying them an Early Cretaceous age, which in the mountain range of the same name, as the upper part of the *Yahgan Fm.,* also represents the continuity of the local regional geology of the Scotia plate beyond the present track of the Magellanes–Fagnano Fault.

In 1932, Kranck mapped the region located north of Lake Fagnano or Lake Khami (a native selknam denomination) (Fig. 3.8), and he concluded that the Sierra Beauvoir (Beauvoir Range) is composed of a Cretaceous flysch (a sedimentary deposit consisting of thin beds of shale alternating with coarser graywacke strata), an important fact since these concepts came from the creator of the *Yahgan Fm.*, then mistakenly considered as of Paleozoic age.

Some other authors have been more sensitive concerning the matter of the Early Cretaceous formations. Thus, Borrello (1967) applied the concept of flysch, taken from Alpine geology to the wide cluster of the *Yahgan–Alvear–Monte Olivia* and *Beauvoir Formations*. It should be important to add to this, in coincidence to what Kranck stated, that the flysch deposition continued until the earliest Tertiary. This criterion of unification is the one followed by Di Benedetto (1973) who understood the Alvear Range as *Yahgan Fm.*, describing darker schists which were compared to the rocks of the Sorondo–Valdivieso Ranges. Then, Yrigoyen (1962) proposed his *Formación Vicuña* (Fig. 3.9) for the western portion of the Beauvoir Range and a smaller sector to the south of Lake Fagnano, of Early Cretaceous times. Harrington (1943) brought a suggestion of Late Jurassic–Early Cretaceous age to the *Serie*

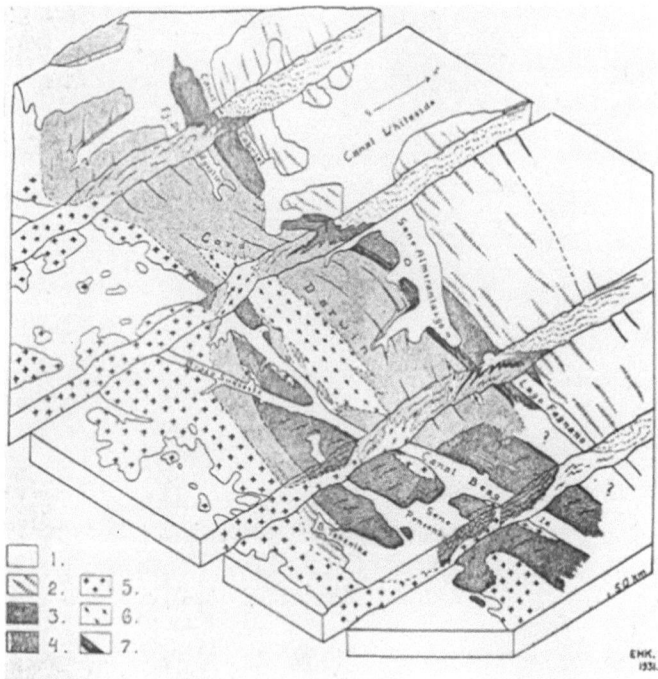

Fig. 3.8 Original tectonic analysis of the central parts of the Andean Cordillera of Tierra del Fuego (Kranck 1932a). 1. Tertiary "Molass" (Magallanean beds). 2. Cretaceous "Flysch" of the Marginal Cordillera. 3. The Yahgan and Monte Buckland formations. 4. The high-grade metamorphic schists of the Central Cordillera. 5. Andean diorite. 6. Central granite. 7. Quartz porphyries

Pizarreña located in northern Isla de los Estados, the same units which Caminos and Nullo (1979) typified as *Beauvoir Formation*.

An extended criterion about the contemporary deposition and assumed identity of the *Yahgan Fm.* and the *Beauvoir Fm.*, which would simply define an Early Cretaceous horizon characteristic of the Patagonian Andean Cordillera. This concept overcomes any concern about formational names different of *Yahgan Formation*, which, undoubtedly, has priority.

It is undeniable that the matter is extremely difficult concerning a formational definition, working with isolated observations or tentatively extended ones, or using interpretative criteria. This is illustrated by recognition in Olivero and Martinioni (2001, p. 181) that the widespread mudstones of the *Beauvoir Fm.* also enclose other units as young as the Paleocene.

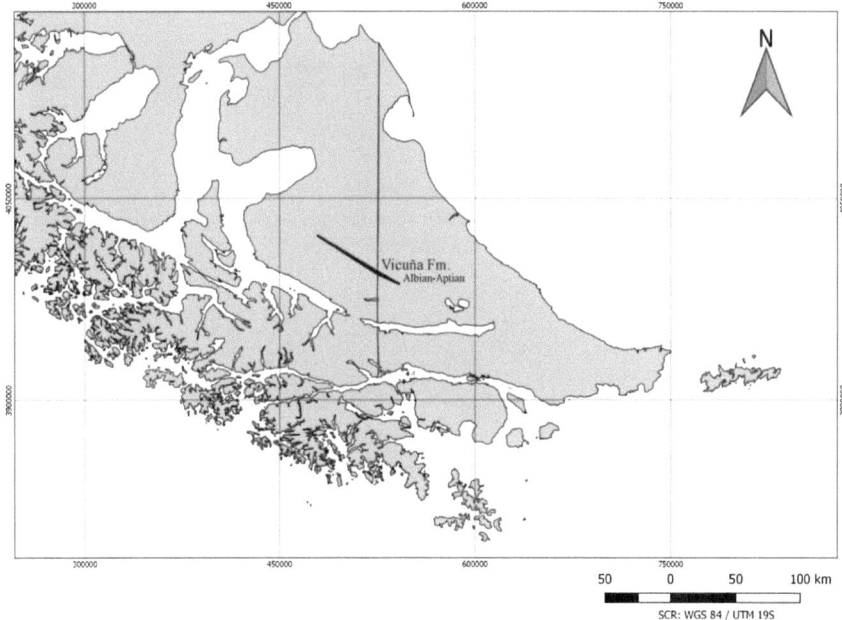

Fig. 3.9 Map of Isla Grande de Tierra del Fuego highlighting the position of the Vicuña Formation (Aptian–Albian age) in Argentina and Chile (modified from Yrigoyen 1962)

3.1.2 The Stratigraphic Unit or "Serie Porfirítica" (Porphyritic Serie) Predating the Tithonian–Neocomian Times or the Yahgan Formation

This case is simpler than the one of the *Yahgan Fm.* concerning its identification, as being the quartz porphyries the key element in the analysis. Some difficulties are found at certain sectors where those porphyries are absent. This higher reliability in the identification of the "*Serie Porfirítica*" comes together with the antiquity of the discovery of the quartz volcanic rocks in Isla de los Estados, both by Lovisato (1883) and Nordenskjöld (1905).

The indicative value of those quartz porphyries was then generalized (Bonarelli 1917, Kranck 1932a) and particularly by Harrington (1943) who named "*Serie Porfirítica*" (Porphyritic Series) the stratigraphic unit which holds them, giving them the Patagonian Cordillera character that involves the Fuegian Andes, adding a detailed reference about the deformed condition of these massive rocks (Fig. 3.10).

Borrello (1969) coined the name of *Lemaire Formation* for the stratigraphic unit composed of porphyry rocks in the Isla de los Estados and south of Lake Fagnano. Caminos (1975) compared and linked these porphyries with those of the *El Quemado Fm.* in Santa Cruz province (southern Patagonia), giving a warning in 1980 that the porphyries are comprised in the *Alvear Fm.* (Petersen 1949), a matter which

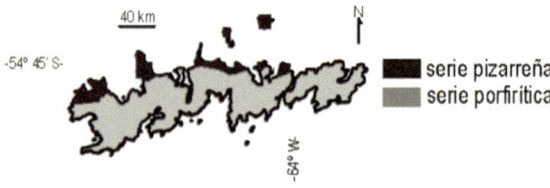

Fig. 3.10 Geologic sketch of Isla de los Estados by Harrington (1943) with original denominations of geological entities ("serie porfirítica" is the Porphyritic series and "serie pizarreña" is the slaty series or *Yahgan Fm.*)

Fig. 3.11 Author's interpretation of the relationship between stratification (S$_0$) and slaty cleavage (S$_1$) in the Yahgan folding areal context. *Credit* R. D. Acevedo

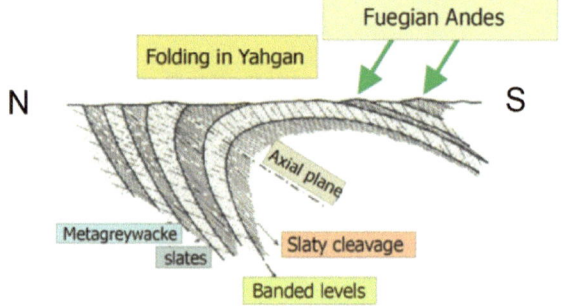

was revalued by Caminos et al. (1981). Thus, it is exposed the important question about the identity or distinction of the *Alvear Fm.* and *the Lemaire Fm.*, besides the fact about the oversighting of the prioritary name of "*Serie Porfirítica*" devised by Harrington (1943).

This last situation could be the subject of criticism as Harrington did not use the concept of *formation*, but that was an historical circumstance which does not reduce the value of the term "*serie,*" well consolidated in Argentine geology. The criticism of the expression "*Porfirítica*" could also be taken as a minor subject, although there is misunderstanding between "*porfirítica*" (porphyritic) or "*andesita pre-terciaria*" (pre-Tertiary andesite) and "*pórfiro cuarcífero*" (quartz porphyry) or "*liparita pre-terciaria*" (pre-Tertiary liparite), since the term "*porfirítica*" or "*porfírica*" is just the result of using the name "*pórfiro*" (porphyry) as an adjective.

This stratigraphic unit, even without having reliable chronopalaeontological data, is a well known, composed sedimentary–volcanic mixture.

The above-mentioned doubt about the outcrops that could belong to the "*Serie Porfirítica*" exists along National Route 3, clarifying the distinction of the *Yahgan Fm.* in Mount Olivia, with the so-called *Alvear Fm.* in the Alvear Range. Indeed, the upper part of the Mount Olivia has been carved in what are clearly *Yahgan Fm.* schists with their southwards dipping folds, since the axial plane cleavage has modeled the shape of Mount Olivia itself (Fig. 3.11).

Over the rim of the Lashifashaj valley alignment, in the northern hillside of the Sorondo Range, the schists are different and the quartz porphyries re-appear. The

Fig. 3.12 Arroyo Rojo site. *Credit* G. Cabrera

contacts between the volcanic rocks and the schists show the wearing of the latter by a process of "assimilative dissemination" of chemical–mineralogical and mechanical origin, in such a way that the schists of tuffaceous appearance are contaminated by the alteration products from the volcanic rocks, as chlorite and titanite, and the volcanic rocks show some schistosity close to the diffuse contact between both units. The central part of the volcanic rocks, that is the part located far away from the contact, maintains its coherence and no alteration is visible. The above-mentioned facts are clear examples of the responses to different strengths by the rigid volcanic rocks and the more elastic or plastic schists. It is a case like one of the contacts between quartz porphyries and schists, which in an extreme case has finished with the vanishing of the volcanic rocks and a compositional change of the schists.

The acid eruptive rocks of the *Lemaire Fm.*, bearing volcanogenic massive sulphide deposits in *Arroyo Rojo* (Red creek) and Laguna Guanaco (Guanaco pondake) are exposed with sheets of *Yahgan Fm.* around Encajonado creek (Fig. 3.12).

In the northern hillsides of the Sorondo Range, the porphyries indicate the presence of the "*Serie Porfirítica,*" in contrast with the southern hillsides that dip toward Canal Beagle, composed of the *Yahgan Fm.*

These last statements allow a reconstruction of a north–south profile (Fig. 3.13). This interpretative essay, which exposes a greater tectonic folding, is compatible with the profile presented by Caminos (1980) which does not enclose the Beauvoir

Fig. 3.13 Structural sketch of the Fuegian Andes at longitude 68° 10′ W (Cerro Rodriguez, 54° 30′ S, 68° 10′ W). Author's interpretation

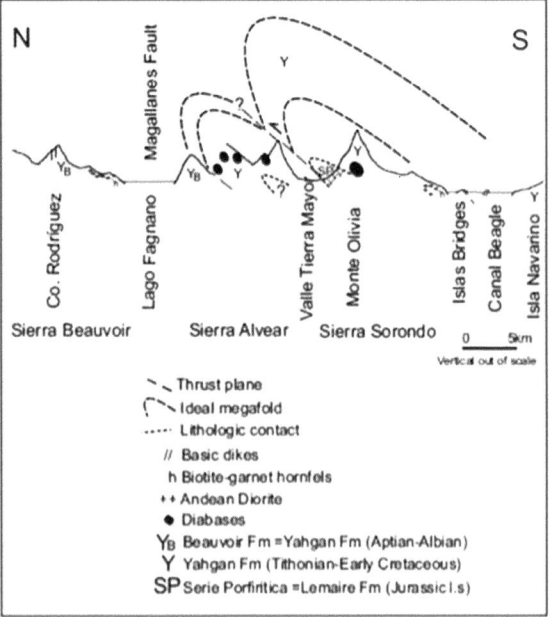

Range but expresses a similarity in the faults or slips of low angle in the Carbajal valley and in Lake Fagnano.

Some critical locations of the deformed and folded "*Serie Porfirítica*" in the Isla Grande de Tierra del Fuego are presented by Acevedo (1986, 1988, 1992b).

Two characteristics of the "*Serie Porfirítica*" must be added. One is the problem concerning its base and roof. Mount Buckland, following Kranck, shows the contacts with both the overlying (*Yahgan*) and the underlying rock formations.

The lower contact is placed over "metamorphic basal schists," where an abnormal superposition is assumed. The upper contact is with beds attributed to the *Yahgan Fm.* The structural complexity makes difficult to reach a safer interpretation. Kranck himself reported some intercalations of quartz porphyries and ftanitic schists in Mount Buckland. This is a valuable statement, since it comes from the creator of the formation. The profile by Kranck of this same site is thus presented as a simplification inside the complex structure. The assumption of an overturned fold, with repetition of the metamorphic rocks and the "*Serie Porfiritica*" in its center, cannot be rejected. Figure 3.13 is an attempt of a possible interpretation of the sequence.

The appreciations by Caminos and Nullo (1979) suggest a transition between the "*Serie Porfirítica*" or *Lemaire* and the *Yahgan Fm.*, which gives a clear example of how complex, vague, and indefinite is this matter. Escayola et al. (2007) contribute isotopic information to the debate.

Those features stated above, together with the complexity resulting from the folding process and the possibility of abnormal contacts, put limits to the formational characteristics as a horizon or as a stage.

Fig. 3.14 Outcrops of metavolcanic rocks from Ensenada Bay crossed by quartz veins. *Credit* R. D. Acevedo

The second aspect is the position and meaning of the *"Serie Porfirítica"* in the stratigraphic series. Firstly, the *"Serie Porfirítica"* is involved in the deformation and folding, as it happens with the superimposed layers. But this fact does not remove the occurrence of the *"Serie Porfirítica"* being the basement on which the Early Cretaceous marine horizons (including the Tithonian, i.e., the beginning of the Andic Cycle) are settled. It could be "basement" in a restricted sense that may be called "Basement One," as it can be seen in the acid metavolcanic rocks (Fig. 3.14), exposed along the coast west of *Bahía Ensenada* (Ensenada Bay) and overlying the real basement of regional schists, superimposed to the true basement or "Basement Zero," which brings us back to the matter whether these are or are not the schists of the Lapataia Formation, as it will be seen later on.

This Neoproterozoic basement of variable extension has been involved in orogenic events prior to the Andean orogeny, active since the Cretaceous (Folguera et al. 2018).

Without any doubts, the so-called porphyries or the *Serie Porfirítica* have a regional sense in both Andean and extra-Andean Patagonia, linked to the matter about the depth reached by the folding deformation and the associated fractures in the orogenic belt. It is probable that the rocks of the Andean formations of the *"Serie Porfirítica"* were alternatively subject of marine and subaerial erosion: The abundance of porphyries suggests a greater instability in the overlying layers, with intermittent rises to subaerial conditions.

References

Acevedo RD (1986) Datos estructurales y litológicos de la porción oriental de Península Mitre, Tierra del Fuego. 3ª Reunión de Microtectónica, Actas: 104–108 (Universidad de La Plata. La Plata)

Acevedo RD (1988). Estudios geológicos areales y petroestructurales en el Complejo Deformado de los Andes Fueguinos. Universidad Nacional de Buenos Aires. Biblioteca de la Facultad de Ciencias Exactas y Naturales. Ph.D. Thesis, 233pp

Acevedo RD (1990) Destape de cuerpos plutónicos ocultos en Península Ushuaia, Tierra del Fuego. 11º Congreso Geológico Argentino, 1:153–156 (San Juan)

Acevedo RD (1992a). Los anfíboles cálcicos como indicadores del origen magmático intrusivo de las rocas melanocráticas del Batolito Andino en Tierra del Fuego. VIIIº Congreso Latinoamericano de Geología, Actas 4:163–167 (Salamanca)

Acevedo RD (1992b) Las rocas eruptivas ácidas del Complejo Deformado de los Andes Fueguinos. Monografía de la Academia Nacional de Ciencias Exactas, Físicas y Naturales, Actas 8:45–48 (Buenos Aires)

Acevedo RD (1996) Los mecanismos sustitutivos y los factores de evolución en los anfíboles de la Hornblendita Ushuaia, Tierra del Fuego. Revista de la Asociación Geológica Argentina, 51(1):69–77 (Buenos Aires)

Acevedo, RD, Linares E, Ostera H, Valín ML (2002) La Hornblendita Ushuaia (Tierra del Fuego): Geoquímica y geocronología. Revista de la Asociación Geológica Argentina, 57 (Buenos Aires)

Acevedo RD, Quartino G, Coto C (1989) La intrusión ultramáfica de Estancia Túnel y el significado de la presencia de biotita y granate en la Isla Grande de Tierra del Fuego. Acta Geológica Lilloana, 17(1):21–36 (San Miguel del Tucumán)

Acevedo RD, Roig CE, Linares E, Ostera HA, Valín-Alberdi ML, Queiroga-Mafra JM (2000). La intrusión plutónica del Cerro Jeu-Jepén. Isla Grande de Tierra del Fuego, República Argentina. Cadernos do Laboratorio Xeolóxico de Laxe, 25:357–359 (A Coruña)

Acevedo RD, Roig CE, Valín-Alberdi ML (2004) Lithologic types of Jeu-Jepén Diorite, Isla Grande de Tierra del Fuego. International Symposium on the Geology and Geophysics of the Southernmost Andes, the Scotia Arc and the Antarctic Peninsula. Bolletino di Geofisica teorica ed applicata. Instituto Nazionale di Oceanografia e di Geofisica Sperimentale 45(2):100–102

Aguirre Urreta B, Suárez MD (1985) Belemnites de una secuencia turbidítica volcanoclástica de la Formación Yahgán – Titoniano-Cretácico inferior del extremo Sur de Chile. IV Congreso Geológico Chileno, Actas I(1):1–16

Biel C, Subías I, Fanlo I, Acevedo RD (2007) Mineralogical characterization of Lemaire and Yahgán Formations, Tierra del Fuego, Argentina GeoSur. Santiago de Chile. Libro de Resúmenes, 23

Bonarelli G (1917) Tierra del Fuego y sus turberas. Anales del Ministerio de Agricultura de la Nación. Sección Geología, Mineralogía y Minería, XII, Nº 3 (Buenos Aires)

Borrello AV (1967) Estado actual del conocimiento geológico del flysch en la Argentina. Revista del Museo de La Plata. Tomo VI. Geología, Nº 44:125–153 (La Plata)

Borrello AV (1969) Los geosinclinales de la Argentina. Dirección Nacional de Geología y Minería, Anales 14 (Buenos Aires)

Camacho HH (1948) Geología del Lago Fagnano o Cami. Universidad Nacional de Buenos Aires. Biblioteca de la Facultad de Ciencias Exactas y Naturales. Ph.D. Thesis, 543

Caminos R (1975) Tobas y pórfiros dinamometamorfizados de la Isla de los Estados, Tierra del Fuego. 6º Congreso Geológico Argentino, 2:9–23 (Bahía Blanca)

Caminos R (1980) Cordillera Fueguina. En Geología Regional Argentina. Academia Nacional de Ciencias. Córdoba, II:1463–1501

Caminos R, Haller M, Lapido O, Lizuain A, Page R, Ramos V (1981) Reconocimiento geológico de los Andes Fueguinos. Territorio Nacional de Tierra del Fuego. 8º Congreso Geológico Argentino, 3:754–786 (San Luis)

Caminos R, Nullo F (1979) Descripción geológica de la Hoja 67e: Isla de los Estados. Servicio Geológico Nacional, Boletín nº 175 (Buenos Aires)

Cenni M (2005) Rilevamento geologico e analisi strutturale del settore centrale della cordigliera delle Ande nella Terra del Fuoco. Tesi sperimentale di Laurea. Universidad de Urbino (Italia). 89pp

Coleman RG (1977) Ophiolites Ancient Oceanic Lithosphere? Minerals and Rocks, vol. 12. Springer, Berlin, Heidelberg, and New York, 229pp

Di Benedetto HJ (1973) Mapa geológico de la Cuenca Austral. Yacimientos Petrolíferos Fiscales. Buenos Aires

Dott RH Jr, Winn RD Jr, De Wit MJ, Bruhn RL (1977) Tectonic and sedimentary significance of cretaceous Tekenika Beds of Tierra del Fuego. Nature 266:620–623

Elsztein C (2004) Geología y evolución del Complejo Intrusivo de la península Ushuaia, Tierra del Fuego. Departamento de Ciencias Geológicas de la Facultad de Ciencias Exactas y Naturales de la Universidad de Buenos Aires. Thesis. 103pp

Escayola M, Acevedo R, González-Guillot M, Pimentel M (2007) Preliminary results of Sm-Nd analysis of Yahgán and Lapataia formations, their provenance. Tierra del Fuego, Argentina. GeoSur. Santiago de Chile. Libro de Resúmenes, 54

Folguera A, Contreras Reyes E, Heredia N, Encinas A, Iannelli S, Oliveros B, Dávila F, Collo G, Giambiagi L, Maksymowicz A, Iglesia Llanos MP, Turienzo M, Naipauer M, Orts D, Litvak V, Alvarez O, Arriagada C (2018) The evolution of the Chilean-Argentinean Andes. Springer Earth Systems Sciences, 564pp

González-Guillot M, Acevedo RD, Escayola M (2010) El gabro Rancho Lata: magmatismo meso-zoico off-axis de la cuenca marginal Rocas Verdes en los Andes Fueguinos de Argentina. Revista Mexicana de Ciencias Geológicas 27(3):431–448

González-Guillot M, Escayola M, Acevedo R, Pimentel M, Seraphim G, Proenza J, Schalamuk I (2009) The Plutón Diorítico Moat: mildly alkaline monzonitic magmatism in the Fuegian Andes of Argentina. J South Am Ear Sci 28(4):345–359

González Guillot M, Ghiglione M, Escayola M, Martins Pimentel M, Mortensen J, Acevedo RD (2018) Ushuaia Pluton: magma diversification, emplacement and relation with regional tectonics in the southernmost Andes. J S Am Earth Sci (submitted)

González-Guillot M, Prezzi C, Acevedo RD, Escayola M (2012) A comparative study of two rear-arc Plutons and implications for the Fuegian Andes tectonic evolution: Mount Kranck Pluton and Jeu-Jepén Monzonite, Argentina. J S Am Earth Sci 38:71–88 (Pergamon-Elsevier Science Ltd., Amsterdam)

González-Guillot M, Urraza I, Acevedo R, Escayola M 2016. Magmatismo básico Jurásico-Cretácico de los Andes Fueguinos y su relación con la Cuenca Marginal Rocas Verdes. Revista de la Asociación Geológica Argentina. Tomo 73(1):1–22

Gregory JW (1915) The Geological relations and some fossils of South Georgia. Trans Royal Soc Edinburgh 50:817–822

Harrington HJ (1943) Observaciones geológicas en la Isla de los Estados. Museo Argentino de Ciencias Naturales. Anales, Tomo XLI. Geología. Publicación N° 29:29–52 (Buenos Aires)

Iovine GM (2005) Estudio microtectónico de la deformación sobreimpuesta al depósito polimetálico Arroyo Rojo, Isla Grande de Tierra del Fuego. Departamento de Ciencias Geológicas de la Facultad de Ciencias Exactas y Naturales de la Universidad de Buenos Aires. Thesis, 91pp

Kranck EH (1932a). Geological investigations in the Cordillera of Tierra del Fuego. Acta Geogr 4(2). 231pp (Helsinki)

Lovisato D (1883) Apuntes geológicos sobre la Isla de los Estados. In Bove G (ed) Expedición Austral Argentina. Instituto Geográfico Argentino. 47–53 (Buenos Aires)

Nordenskjöld O (1905) Die krystallinen gesteine der Magellanslander. Wiss Ergebn Exp Magell 1(6):175240

Olivero EB, Martinioni DR (1996) Late Albian inoceramid bivalves from the Andes of Tierra del Fuego. Age implications for the closure of the Cretaceous marginal basin. J Paleontol 70:272–274

Olivero EB, Martinioni DR (2001) A review of the geology of the Argentinian Fuegian Andes. J S Am Earth Sci 14:175–188

Olivero EB, Martinioni DR, Malumián N, Palamarczuk S (1999) Bosquejo geológico de la Isla Grande de Tierra del Fuego, Argentina. 14° Congreso Geológico Argentino, 1:291–294 (Salta)

Petersen CS (1949) Informe sobre los trabajos de relevamiento geológico efectuados en Tierra del Fuego entre 1945 y 1948. Dirección General de Industria y Minería. Buenos Aires

Richter M (1925) Beiträge zur Kenntnis der Kreide in Feuerland. Neues Jahrbuch fur Geologic und Palaontologie 52:524–568

Steinmann G (1927) Die ophiolitshen zonen in den mediterranen Kettengebirgen, Translated and printed by Bernoulli & Friedman, en Dilek & Newcomb (eds). Ophiolite Concept and the Evolution of Geologic Thought. Geological Society of America Special Publication 373:77–91

Tyrrell GW (1930) The petrography and geology of South Georgia. Report on the geological collections made during the voyage of the "Quest" on the Shackleton-Rowett expedition 1921–1922, 1–24

Villar L, Acevedo RD, Lagorio S (2007) The Puente Quemado Gabbro, to the west of Ushuaia, Tierra del Fuego, Argentina. GeoSur. Santiago de Chile. Libro de Resúmenes, 172

Wilckens O (1930) Fossilien und Gesteine von Süd-Georgien. Scientific Results of the Norwegian Antarctic Expeditions. 1927–1928 and 1928–1929, 8, 26pp

Yrigoyen MR (1962) Evolución de la exploración petrolera en Tierra del Fuego. Petrotécnica (Instituto Argentino del Petróleo), Anales XII, N° 4:28–38

Chapter 4
The Basic Eruptive Rocks of Alvear and Sorondo Ranges (Mount Olivia)

Another matter of interest is the one concerning the basalts–andesites and diabase–microgabbros of the Alvear and Sorondo ranges (Fig. 4.1). They have been described as vein-layer interbedding in the sedimentary frame of the Yahgan Fm. (Caminos et al. 1981; Quartino et al. 1987, 1989) and were joined later to the Lemaire Fm. as an expression of bimodal volcanism (Ametrano et al. 1999; Olivero et al. 1999) to explain the mineralization model of the volcanogenic massive sulfides.

At the lower sector of Mount Olivia, cut by the road, a massive diabase (or microgabbro) with spilitic features has intruded slates. In Paso Garibaldi (next to the panoramic point), undistinguishable processes have taken place. Other similar eruptive intercalations are exposed in the section of the road that crosses the Alvear Range, up to the site of Rancho Hambre in the sector of National Route 3 that climbs to Paso Garibaldi, next to Lake Escondido (Fig. 4.2).

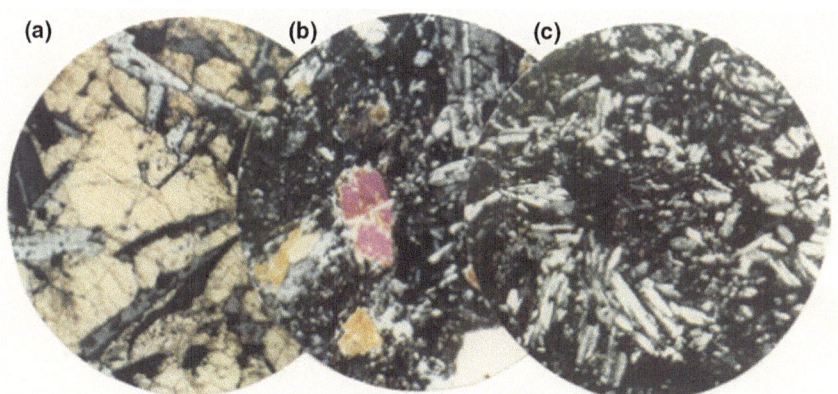

Fig. 4.1 **a** Ophitic, **b** intergranular, and **c** intersertal textures in diabases from the Alvear and Sorondo ranges, magnification ×60, XPL. *Credit* R. D. Acevedo

© The Author(s), under exclusive licence to Springer Nature Switzerland AG 2019
R. D. Acevedo, *Geological Records of the Fuegian Andes Deformed Complex Framed in a Patagonian Orogenic Belt Regional Context*, SpringerBriefs in Earth System Sciences, https://doi.org/10.1007/978-3-030-00166-7_4

Fig. 4.2 Interfingering between eruptive and sedimentary levels with dynamic schistosity, Paso Garibaldi (54° 38′S, 67° 45′W). *Credit* R. D. Acevedo

Fig. 4.3 Lenses of subvolcanic bodies are exposed along the geological section through the Alvear Range. *Credit* M. Menichetti

These rocks are lenses of leucobasalts and melandesites (Fig. 4.3) which were the subject of strong deformation (Quartino et al. 1987, 1989; Acevedo 1988; Cenni 2005; González-Guillot et al. 2016) and schist from the core to the borders.

References

Acevedo RD (1988) Estudios geológicos areales y petroestructurales en el Complejo Deformado de los Andes Fueguinos. Universidad Nacional de Buenos Aires. Biblioteca de la Facultad de Ciencias Exactas y Naturales. Ph. Thesis, 233p

Ametrano S, Etcheverry R, Echeveste H, Godeas M, Zubía M (1999) Depósitos polimetálicos (tipo VMS) en la cordillera fueguina, Tierra del Fuego. In: Zapettini EO (ed) Recursos Minerales de la República Argentina. Buenos Aires, pp 1029–1038

Caminos R, Haller M, Lapido O, Lizuain A, Page R, Ramos V (1981) Reconocimiento geológico de los Andes Fueguinos. Territorio Nacional de Tierra del Fuego. 8° Congreso Geológico Argentino 3:754–786 (San Luis)

Cenni M (2005) Rilevamento geologico e analisi strutturale del settore centrale della cordigliera delle Ande nella Terra del Fuoco. Tesi sperimentale di Laurea. Universidad de Urbino (Italia), 89p

González-Guillot M, Urraza I, Acevedo R, Escayola M (2016) Magmatismo básico Jurásico-Cretácico de los Andes Fueguinos y su relación con la Cuenca Marginal Rocas Verdes. Revista de la Asociación Geológica Argentina. Tomo 73(1):1–22

Olivero EB, Martinioni DR, Malumián N, Palamarczuk S (1999) Bosquejo geológico de la Isla Grande de Tierra del Fuego, Argentina. 14° Congreso Geológico Argentino 1:291–294 (Salta)

Quartino BJ, Acevedo RD, Scalabrini Ortiz J (1987) Rocas eruptivas vulcanógenas entre Monte Olivia y Paso Garibaldi, Isla Grande de Tierra del Fuego. Simposio Internacional de Vulcanismo Andino, Actas IV: 209–212 (San Miguel del Tucumán)

Quartino BJ, Acevedo RD, Scalabrini Ortiz J (1989) Rocas eruptivas volcanógenas entre Monte Olivia y Paso Garibaldi, Isla Grande de Tierra del Fuego. Revista de la Asociación Geológica Argentina 44(3–4):328–335 (Buenos Aires)

Chapter 5
The Cretaceous Layers Later than the Early Cretaceous

The marine continuity after the Early Cretaceous implies an essential difference with the condition in the orogenic belt with a N-S trend of the Patagonian Cordillera, and it could be said, in a figurative sense, that the continent has kept its feet in the ocean. But this important difference does not avoid the matter concerning if these Late Cretaceous layers have been in fact involved in the same folding complexity as the Early Cretaceous ones, with slight metamorphism. The available information results, at present, not complete enough to adjust a formational definition.

The known formations (Olivero 2002; Olivero et al. 2002) state clearly the existence of Late Cretaceous marine layers, but they keep the difficulties of their identification in the field, like, e.g., the case about the separation of the Early and Late Cretaceous mudstones (Olivero et al. 2002). To this, Martinioni (1997) added that the mudstones of the Beauvoir Fm., to say Yahgan Fm., or Cretaceous units, comprise rocks as young as the Paleocene. Thus, so far, it results very doubtful to make any final characterization of the formations coming from a lithologic base only, lacking also precise paleontological data.

Acevedo (1988) showed the regional presence, enclosed in pelitic beds attributed to the Fuegian Andes Deformed Complex, of *Inoceramus steinmanni* and *Inoceramus* sp. (identified by H. H. Camacho, personal communication) in the Montes Negros, and also *Inoceramus* sp. (identified by F. Medina, personal communication) in the eastern side of Bahía Aguirre (Acevedo 1988, p. 87) reproduced here in Figs. 5.1, 5.2, 5.3, 5.4 and 5.5, that were later the subject of attention in Olivero and Medina (2001).

Fig. 5.1 *Inoceramus steinmanni* Wilckens (Upper Cretaceous) ×1 (Montes Negros). *Credit* R. D. Acevedo

Fig. 5.2 Inoceramid undetermined ×1 (Buen Suceso bay). *Credit* R. D. Acevedo

Fig. 5.3 Inoceramid undetermined ×0.3 (San Mauricio cove). *Credit* R. D. Acevedo

Fig. 5.4 *Inoceramus* sp. (external mold) ×1 (Montes Negros). Cfr. *Inoceramus* sp. In Thomson et al. (1982). *Credit* R. D. Acevedo

Fig. 5.5 Inoceramid
undetermined ×0.3 (Montes
Negros). *Credit* R. D.
Acevedo

References

Acevedo RD (1988) Estudios geológicos areales y petroestructurales en el Complejo Deformado de los Andes Fueguinos. Universidad Nacional de Buenos Aires. Biblioteca de la Facultad de Ciencias Exactas y Naturales. Ph.D. Thesis. 233pp

Martinioni DR (1997) Cretaceous-Paleogene surface Stratigraphy of the Austral Basin in the southernmost Andes. New evidences from the central Tierra del Fuego, Argentina. Gaea Heidelb 3:231–232

Olivero EB (2002) Petrografía sedimentaria de sistemas turbidíticos del Cretácico-Paleógeno, Andes Fueguinos: procedencia, volcanismo y deformación. 15° Congreso Geológico Argentino. El Calafate, Santa Cruz. CD-ROM, Artículo n° 015. 2p

Olivero EB, Medina FA (2001) Geología y paleontología del Cretácico marino en el sureste de los Andes Fueguinos, Argentina. Rev de la Asoc Geol Argent 56(3):344–352. Buenos Aires

Olivero EB, Malumián N, Palamarczuk S, Scasso RA (2002) El Cretácico superior-Paleógeno del área del Río Bueno, costa atlántica de la isla Grande de Tierra del Fuego. Rev de la Asoc Geol Argent 57(3):199–218. Buenos Aires

Thomson MRA, Tanner PWG, Rex DC (1982) Fossil and radiometric evidence for ages of deposition and metamorphism of sedimentary sequences on South Georgia. In: Craddock C (ed) Antarctic geoscience, University of Wisconsin Press, Madison, pp 177–184

Chapter 6
The Estimated Basement of the Fuegian Andes Mesozoic Stratigraphic Series and Those Rocks Regarded as Pre-mesozoic in Age

It was already pointed out the possible role of the "*Serie Porfirítica*" as basement, as a consequence of the regional deployment of the quartz porphyries in Patagonia. This would be a restricted concept of the word basement, as stated before.

Besides this aspect, it is of interest to note the lepto-metamorphic rocks seen on the coastal area of the Parque Nacional Tierra del Fuego (Tierra del Fuego National Park) and neighboring areas. They could be regarded as the basement of the Mesozoic marine column (Fig. 6.1).

These rocks have been called by several different names: Kranck in 1932 included those observed in Chile, in the Darwin Cordillera, within the generalized concept of high metamorphic schists, as metamorphic rocks of medium to high grade. Petersen (1949) simply called then "*Esquistos de Lapataia*" (Lapataia's schists); Borrello (1969) called them "*Metamorfita Lapataia*," and Olivero et al. (1999) coined the name *Lapataia Formation*, to finally name them as "*basamento*" (basement) in Olivero and Martinioni (2001). In only one opportunity, the concept of formation is mentioned, but remarking the possibility of either Paleozoic or Mesozoic age. This last discussion clearly shows that the matter is, in fact, still an open one.

The fact is that these folded and metamorphosed rocks, belonging to a greenschist facies (Fig. 6.2), are isolated from the ones of "*Serie Porfirítica*" or in abnormal contact. An exception is the one remarked by Kranck at Mount Buckland in page 61, where the "*Serie Porfirítica*" is lying on the "*esquistos basales*" (basal schists), probably correlated with the rocks of Lapataia Bay. But, as it was mentioned before, even this unusual case of a possible contact between the "*Serie Porfirítica*" and the probable metamorphic basement is doubtful, regarding the existence of an overturned fold in such a way that the basal schists belong to the *Yahgan Fm.*, by repetition.

Concerning the metamorphic grade of the Lapataia rocks, it is indeed intermediate between the one of Darwin Cordillera and those of the Mesozoic series, i.e., *Yahgan Fm.* In these criteria, the metamorphic grade decreases eastward from Darwin Cordillera to Lapataia. The subject, of petrologic nature, deserves a more detailed study. It should be kept in mind the existence of a contact metamorphism factor on

© The Author(s), under exclusive licence to Springer Nature Switzerland AG 2019
R. D. Acevedo, *Geological Records of the Fuegian Andes Deformed Complex Framed in a Patagonian Orogenic Belt Regional Context*, SpringerBriefs in Earth System Sciences, https://doi.org/10.1007/978-3-030-00166-7_6

Fig. 6.1 Local basement characterized by schistosity dipping to the SSW at a beach in the Tierra del Fuego National Park. *Credit* M. Cenni

Fig. 6.2 Contorted outcrops of plastic schists from Lapataia Bay. *Credit* R. D. Acevedo

the metamorphic schists near Estancia Túnel, where both garnet and biotite were generated (Acevedo et al. 1989).

Relating to the distinction between the Lapataia schists and the folded and metamorphosed Mesozoic group, three features have been used in the literature:

The first is the metamorphic grade, low in the greenschist facies but greater in the *Yahgan* schists and in the "*Serie Porfirítica.*" They could be classified as leptogneiss, i.e., a low temperature gneiss, without potassium feldspar (Fig. 6.3). But it must be kept in mind that in a lepto-metamorphosed and strongly folded complex, it is foreseeable a gradual metamorphism variation by the irregular availability of oxhidriles

Fig. 6.3 Regional basement of leptogneisses in the Acigami Lake. *Credit* R. D. Acevedo

and a possible non-homogeneous thermal behavior. Particularly, the conditions in the deepest root areas of the acutely folded complex are unknown and a higher metamorphism could be in fact present there. In any way, the difference in metamorphic grade between the Mesozoic complex and the Lapataia rocks is very slight and the presence of amphibolites in the last one (Caminos 1980) supports a low grade of metamorphism because that is in fact a tremolite caused by the metamorphism of a former basic constituent. In the same sense, the presence of garnet (Olivero et al. 1997) could be signal of metamorphic contact of a possible, unexposed intrusive body as it occurred in the Ushuaia Peninsula (Acevedo 1990; Elsztein 2004).

The second fact is the much complex folding in detail, consisting of refolding in the schistosity surfaces on Acigami Lake shores (Fig. 6.4). This crenulation is probably the most characteristic pattern or feature of the Lapataia schists pointing toward a second event of deformation. Many details and features of dynamic metamorphism origin are present in the Mesozoic schists (Acevedo 1987; Quartino et al. 1988) in such a way that their place in the metamorphic anticlinorium is again a riddle.

The third fact is the existence of quartz veins and lenses well noticeable in the coasts of Canal Beagle (Fig. 6.5), in relation with the last pattern, to say, quartz veins in the microfolds. Again, this is not a critical quality. It is really a matter of distribution inside a liquid phase, like it is the case in the neighboring areas between a slightly venous micacites and gneisses, a frequent case in the basement of Sierras Pampeanas (Pampean Ranges). This is not a whimsical observation because gneisses of metamorphic grade 2 or 3, with quartz–feldspar veins, are homologous to the fluidal venous schists with quartz veins in metamorphic grade one, e.g., those seen in Lapataia. On the other hand, quartz veins are frequent in the Mesozoic Fuegian schists, in such a way that the most characteristic feature is still the second one that is the refolding, although being still unconclusive.

Fig. 6.4 Refolding in the basement rocks. *Credit* R. D. Acevedo

Fig. 6.5 Quartz-bearing leptogneiss and details of efflorescences of cooper carbonates. *Credit* R. D. Acevedo

Moreover, some schists showed in maps as Lapataia Formation are indeed liparites of Lemaire Formation strongly milonitized levels. Cao et al. (2018) extended this assertion to the entire Lapataia area.

It should be finally noted, making a comparison with another district in the Patagonian Cordillera, e.g., at the Fontana Lake, Chubut Province, that there are isolated lepto-metamorphic rocks of Arroyo Pedregoso, undoubtedly the base of the Mesozoic column, because of the lack of evidence of deformation at that site. The case of Tierra del Fuego is far more complex or doubtful as the metamorphism, although slightly, and the folding deformation disturbed the entire group or complex.

References

Acevedo RD (1987) Río Pipo: una localidad crítica en el estudio estructural de los Andes Fueguinos de la Isla Grande de Tierra del Fuego. 4a Reunión de Microtectónica, Actas. Universidad de San Juan. San Juan, pp 1–5

Acevedo RD (1990) Destape de cuerpos plutónicos ocultos en Península Ushuaia, Tierra del Fuego. 11° Congreso Geológico Argentino, 1: 153–156. San Juan

Acevedo RD, Quartino G, Coto C (1989) La intrusión ultramáfica de Estancia Túnel y el significado de la presencia de biotita y granate en la Isla Grande de Tierra del Fuego. Acta Geol Lilloana 17(1):21–36. San Miguel del Tucumán

Borrello AV (1969) Los geosinclinales de la Argentina. Dirección Nacional de Geología y Minería, Anales 14. Buenos Aires

Caminos R (1980) Cordillera Fueguina. Geología Regional Argentina. Academia Nacional de Ciencias. Córdoba, II, pp 1463–1501

Cao S, Torres Carbonell PJ, Dimieri LV (2018) Structural and petrographic constraints on the stratigraphy of the Lapataia Formation, with implications for the tectonic evolution of the Fuegian Andes. J S Am Earth Sci 84:223–241

Elsztein C (2004) Geología y evolución del Complejo Intrusivo de la península Ushuaia, Tierra del Fuego. Departamento de Ciencias Geológicas de la Facultad de Ciencias Exactas y Naturales de la Universidad de Buenos Aires. Thesis. 103pp

Olivero EB, Martinioni DR (2001) A review of the geology of the Argentinian Fuegian Andes. J S Am Earth Sci 14:175–188

Olivero EB, Acevedo RD, Martinioni DR (1997) Geología y estructura del Mesozoico de Bahía Ensenada, Tierra del Fuego. Rev de la Asoc Geol Argent Tomo 52(2):169–179

Olivero EB, Martinioni DR, Malumián N, Palamarczuk S (1999) Bosquejo geológico de la Isla Grande de Tierra del Fuego, Argentina. 14° Congreso Geológico Argentino, 1: 291–294. Salta

Petersen CS (1949) Informe sobre los trabajos de relevamiento geológico efectuados en Tierra del Fuego entre 1945 y 1948. Dirección General de Industria y Minería, Buenos Aires

Quartino BJ, Acevedo RD, Benito J, Quartino G (1988) Algunas micro y mesoestructuras en rocas leptometamórficas del Complejo Deformado de los Andes Fueguinos. 5ª Reunión de Microtectónica. Actas Universidad de Córdoba. Córdoba

Chapter 7
Eastern Sector of the Fuegian Andes Deformed Complex (FADC-ES)

7.1 Lithology and Structures

The FADC-ES comprises the territories located east of an imaginary line projected from Bahía Aguirre to Policarpo Cove, known as Península Mitre (Fig. 7.1). The Montes Negros and Atocha, Campana and Pirámide hills are situated there, at the southern coast, and the Arriola, Real, and Bilbao hills at the northern coast. Beyond the Straits of Le Maire, the Isla Grande de Tierra del Fuego is separated from Isla de los Estados, whose geologic continuity with the previous locality allows its inclusion as part of the Eastern Deformed Complex.

7.2 Background

There are few published contributions about this region, which are usually difficult to access. Although Bonarelli (1917) and Kranck (1932) included in their works separate sketches where the sector appears mapped, it was Harrington (1943) who, from Isla de los Estados, made the first regional observations, and Caminos and Nullo (1979) compiled the corresponding geological sheet. Furque and Camacho (1949) and Furque (1966) completed, already in Península Mitre, the expeditious knowledge of the area. Later, Caminos et al. (1981) addressed the formational nomenclature in a comprehensive study of the Fuegian Andes. From 1984 onward, the Museo Territorial de Ushuaia initiated scientific expeditions, as part of a pioneer program, of greater scope, which involved the multidisciplinary geological study of the region. Since then, this remote region of the Isla Grande de Tierra del Fuego has been the object of expeditions motivated by the stimulus of the remote and unreachable.

© The Author(s), under exclusive licence to Springer Nature Switzerland AG 2019
R. D. Acevedo, *Geological Records of the Fuegian Andes Deformed Complex Framed in a Patagonian Orogenic Belt Regional Context*, SpringerBriefs in Earth System Sciences, https://doi.org/10.1007/978-3-030-00166-7_7

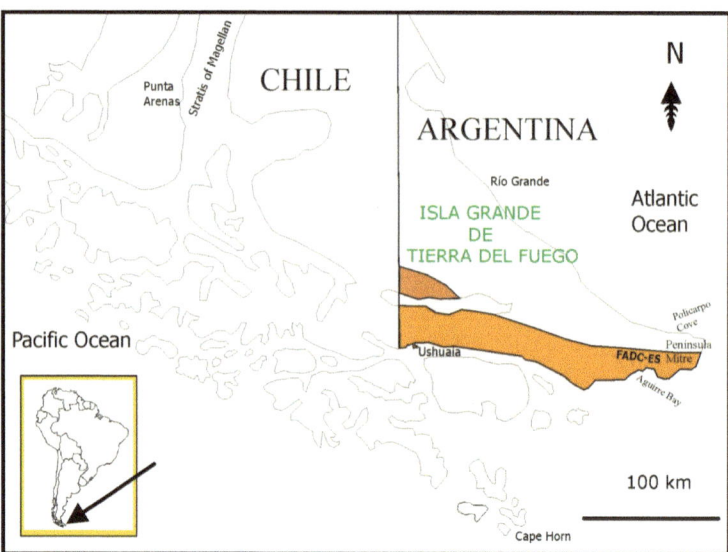

Fig. 7.1 Geographical position of Península Mitre

7.3   References to the Areal Geology

The Deformed Complex of the Fuegian Andes in its eastern sector presents a clearer stratigraphic richness than in the western sector previously described in very significant correspondence with the geological features of Isla de los Estados. The main innovations with respect to the western sector are found in the abundance of granitic and granodioritic porphyres in what could be called the southern lithological set and the presence of rich fossiliferous layers of Upper Cretaceous age in the northern lithological set extending from the average latitude of the Montes Negros to Bahía Tethis. The first of the lithological assemblages corresponds to the Lemaire Formation (Borrello 1969; Caminos et al. 1981) or "porphyric series" of Harrington (1943), and in Montes Negros, it covers the southern half of the area. The second of the lithological assemblages, which includes the "Bahía Tethis strata," would correspond according to Furque (1966) to the Beauvoir Formation, that is, to say the "slaty series" that Harrington (1943) identified at Isla de los Estados. The importance of the sector lies therefore in the neighborhood relations of these two lithological groups, with the lower or southern lithological group proposing the question of its comparison with the Yahgan Formation existing in the western sector.

7.4 Detailed Geological Description

This review, according to localities, will be made from the division of the eastern sector of the complex into three units, (1) southern coast (from Bahía Aguirre to Bahía Valentín, with a reference to Bahía Sloggett, and Atocha, Campana, and Pirámide hills), (2) eastern coast (Montes Negros and western coast of the Straits of Le Maire, from Bahía Buen Suceso to Cabo San Diego and references to Isla de los Estados), and (3) northern coast (from Bahía Tethis to Policarpo Cove).

7.4.1 Southern Coast of Península Mitre (from Bahía Aguirre to Bahía Valentín, with a Reference to Bahía Sloggett, and the Atocha, Campana, and Pirámide Hills)

It is described here as the southern coast of Península Mitre, the coastal strip on the Argentine Sea (southern Atlantic Ocean) between the Sloggett and Valentín bays. In this region, two rocky entities crop out, basically represented by granitic–granodioritic porphyres and their box of low-grade metasedimentary rocks that originally were mudstones and sandstones.

7.4.1.1 Bahía Sloggett

The cliffs of Bahía Sloggett are lithologically represented by strata of clear sandstones, interspersed with coal beds. These layers, that were mentioned by Andersson (1906) and identified as cyclothems by Caminos et al. (1981), were designated as the only Tertiary outcrops of the southern coast of Tierra del Fuego.

Separated from the previous geological context, porphyric rock exposures east of Río López were observed. Under the microscope, a rhyolitic–rhyodacitic rock composed of phenocrysts (35%) of quartz, plagioclase, and alkali feldspar may be seen, included in an acid paste (65%) of granophyric texture, grading locally to micrograin, micrographic, and even spherulitic texture. The phenocrysts, from 1 to 2 mm in section, have their edges corroded and appear engulfed by the paste, a characteristic that is highlighted in quartz individuals. The alkali feldspar—proportionally equivalent to quartz—shows patches of twinned albite on an orthose base. Plagioclase, the less abundant mineral among the phenocrysts, is also albite, with polysynthetic and hypidiomorphic twinning. Some residual phenocrysts of biotite have been completely altered to chlorite and opaque minerals. In the paste, pleochroic and birefringent chlorites are preserved, oriented by fluidity. Other minerals are apatite, zircon, titanite, epidote, and iron oxides. The rock can thus be classified as sodium-rich rhyolitic porphyry or as a quartziferous keratophyre.

7.4.1.2   Bahía Aguirre

In the cove of Puerto Español, located to the west of Bahía Aguirre, there are outcrops of sedimentary-banded metamorphic lepto-pelitic-psammitic rocks.

Starting from the abandoned sawmill of Puerto Español toward the southeast, the first appearance of schistose rocks exposed on a wall that constitutes a friction mirror of reduced dimensions, with abundant clay alteration in front, can be observed. A mimetic cleavage can be seen according to the stratification. The schists are accompanied, parallel to its structure, by some quartz veins and, occasionally, by nodules of jaspery material.

From this outcrop to Point Jalón, the sequence appears very deformed and crossed unconformably by quartz injections. The attitude of the structure changes locally from NW to NE, and a corrugation cleavage modified in turn by the schistosity of chlorites and carbonates in parallel alignment can be recognized. Calcareous levels are interbedded here, where the carbonates participate intimately in the composition of the shales or, secondarily, constituting more or less concordant veins. The rocks are fine sandy schists with relictic clasts (approximately 15%) of up to 0.5 mm in diameter, with 90% of quartz fragments "floating" on a calcareous base. The folding, exceeding the plasticity limit, has produced fracturing according to N 55°E and N 05°E strike planes. Regarding the disseminated sulfide mineralization, an increase of the pyrite content toward the southeast is noticed. The pyrite crystals are very thin, although some reach 1 mm^3.

Alteration haloes in Point Jalón are clearly recognizable. There is a quartz mass of approximately 6 m in diameter, having developed crystals of muscovite and phenomena of silicification and sericitization in the country rock composed of slaty schists. Angular quartz extinctions and micas flexures mark the cataclasis that impacted the invasive quartz mass and its country rock schist.

Further on, toward the south, there is a small entrance where the Garibaldi Ranch is located, and about 150–200 m toward the SE, the Gardiner Cove is located. This point on the coast of Bahía Aguirre has turned out to be significant due to the extreme pyritization in certain levels of the banded sequence (Fig. 7.2). On the coast of Gardiner Cove, there are two clastic fractions, ranging from black carbonaceous pellets to light and dark gray psammitic shales, a set whose appearance resembles the turbiditic deposits of Ushuaia with the addition of the interbedding of phtanitic nodules between the sandy and clayey silt alternation.

The concentrations of pyrite appear grouped in limbs and hinges of the smaller folds, being then conformable and of thickness no greater than 2 cm. Pyrite has probably been syngenetically deposited at certain levels under highly reducing conditions, of the euxinic type. However, some pyritic individuals are associated with the numerous hydrothermal quartz veins and guides, so their syntectonic mobilization and relocation are inferred.

From Point Jalón to Gardiner Cove, and even more to the south, abundant pyrite is seen disseminated in the fine facies of the banded sequence.

Levels containing pyrite concentrates appear both north and south of Gardiner Cove, even followed by a courtship of pyrite individuals scattered throughout the

Fig. 7.2 Piritiferous plots of the banded sequence of Gardiner Cove. Quartz veins, piritiferous also, are found parallel to the slaty cleavage. *Credit* R. D. Acevedo

sequence. The length of this outcrop totals approximately 50 m in the NNW direction. The shape of the mineralized appearance is oblong, with an approximate thickness of 2 m at its ends up to 3 m.

Silicification has been intense. There are veins and numerous quartz veins, conformable or oblique, of one or more generations, with crystalline individuals and reniform accumulations of colloidal silica. These veins are probably witnesses to the redistribution process of pyrite.

The axes and structural planes are arranged regularly according to an ENE strike, an attitude that is maintained until the fold of the beacon located about 120 m NW of Point Pique. In Point Pique, the strike is E-W. The rocks are metasandstones and poorly schistose-banded metamudstones, with weak axial plane cleavage.

The coasts of Point Pique show flattened folds, some of them with very sharp hinges. Folding and fracturing reach a high intensity there (Fig. 7.3). At the beach located to the SW, the contact between the metasedimentary rocks that are being described and rocks of domed erosion forms halfway to Point Lobos is perceived.

A special petrographic variety is found near Puerto Español. It could be defined as a "hybrid" rock, of coarse-grained texture, of a somewhat fluid character, composed of quartz, feldspars, abundant muscovite, and biotite very altered to chlorite and zircon. It is interpreted as a zone of intense mixture, in a sector of brecciation of the intrusive magma that formed the masses of the rhyolitic porphyry with the country rock.

On the stretch from Río Bonpland to the Elizalde Beacon, from the left bank of the river to the vicinity of the beacon and islet Elizalde, the lithological characteristics of the lepto-metamorphic sequence are maintained, but the structural attitude of the layers changes. They are tube-shaped problematics bearing (detectable with scanning electron microscope) banded metasedimentary rocks, with slaty cleavage

Fig. 7.3 Tight folds close to the contact between metasedimentary rocks and eruptive bodies. West of Bahía Aguirre. *Credit* R. D. Acevedo

and calcareous facies. Structural planes dipping to the N, limbs and hinges (marked by sandy facies) are folded in turn in other orders according to the previous attitude.

In the stretch from Río Fatiga to Río Mejillón, in the Lasala inlet, two streams dump their waters—the Fatiga and the Tropa rivers—whose flow follows an approximate N-S course. Between one river and the other, the rocks keep the banded aspect, with cleavage of axial plane (strike N 45°W, dip 20°–25°NE). Toward the mouth of the Tropa River, the structure is arranged according to a NE strike, with a folding axis plunging in that direction.

The structure changes again later toward the E, adopting an E-W position between the Tropa and Mejillón rivers, dipping to the north.

In the floodplain of the Tropa and Mejillón rivers, pumicite gravels were observed, not being able to establish if they have been transported by rivers or taken there by the sea, although this last possibility appears as more probable since outcrops of these are unknown rocks in the inner portion of the island.

At the level of Río Mejillón, the structure returns to its normal NE position. The lithology is of mudstones and sandstones with banks of 70–80 cm of thickness. These rocks are silicified and discordantly traversed by quartz veins. There is a noteworthy mylonitization of mudstone levels. Likewise, certain sectors show oxidation of disseminated sulfides. The structure is of folds lying in a regional attitude.

On the left bank of Río Mejillón River—some 150 m before its mouth reaches the sea—the dune sands contain levels of ochers (pigments) of yellowish-red colors. Its presence is attributed to the alteration by weathering of the abundant pyrite accumulated in those levels.

On the open coasts of the Caleta Oriental, the Río Bolsa yields its waters to the sea. Next to its mouth, on the left margin, metasedimentary-banded units are arranged in thin layers, dark mudstone levels, and other silicified clearer silty–sandy beds. The set is strongly folded.

Pyrite is present in these rocks in an interesting proportion, despite the chaotic, unconformable injection of quartz that has removed it. There were, apparently, three events of silicification: (a) the one that impacted on the mudstone later, (b) another one of the conformable veins, and (c) the last of unconformable guides.

The structure is represented by a tight folding, subhorizontal axial planes, several systems of diaclases, and faults and axial plane cleavage (with crystallization of chlorites and sericite).

The cliffs that follow toward the south are favorable for geological observations since on its perpendicular cut to the structure it allows to draw a good transversal profile. The rocks are made up of carbonaceous mudstones, sandstones, silicified sandstones, mylonites, and the occasional, sporadic presence of some highly calcareous mudstone. In the finer grained samples, the development of two cleavages of clays very angular to each other has been observed under the microscope. The development of sericitic fibrils between the microclasts of quartz and plagioclase has also been noted. Another characteristic feature is the chlorite alteration and the epidotization of the superficial levels. The calcareous mudstones are of great significance because remains of inoceramids have been found in them. In thin section, microfossils with no diagnostic age value have also been recognized and plates with calcite crystals perpendicular to the walls, probably of organic origin.

Another site with alteration haloes is located to the SE of the mouth of the Río Bolsa. A high dissemination of pyrite is found in relation to the ENE structure probably modified by hydrothermal stages linked to porphyritic rocks located east of the regional fissure that hosted the subvolcanic manifestations of the Campana and Atocha mountains. There is an intense zone of sulfide oxidation in the locality with numerous quartz veins associated with a general process of silicification. The pyrite crystals have grown to a size of 2–3 mm, cutting the stratification and schistosity planes.

The abrupt geographic morphological features have not allowed the terrestrial access to the eruptive rocks which outcrop toward the SSE. Despite this, samples from a fallen block could be collected that provided interesting petrographic cuts of a rhyodacite porphyry with a hialoporphyritic texture, with quartz phenocrysts and plagioclase (up to 2 mm in section) dipped in a glassy paste of colloform texture.

The quite large (at least for these latitudes) Río Bolsa (or Río de la Vega) drains through a great regional arrangement of NE direction. Its course uncovered some rocky outcrops covered by a peaty soil and dense forest a few meters from the margins. When tracing its course, two critical locations were observed regarding pyritical appearances. In the lower-middle course, there is a small outcrop represented by black slates, deeply jointed, rusted, semi-separated by soil and forest on the right bank of the river. There was abundant disseminated mineralization of pyrite there, in a smaller, tight, and retracted structure, with a general heading N 15°W, dipping 48° to the NE, which is almost transverse to the river course in one of its bends. The pyrite crystals reach sizes of up to 1 and 2 mm^3 and cut the planes of stratification and cleavage of the mudstones. From the previous point to the next locality (as regarding the anomalous appearance of pyrite), a fine spread of the sulfide is still observed on the river bed boulders, composed of the two clastic fractions of the metasedimentary rocks. On the same bed, the local structure can be seen, of NE-SW orientation, whereas the inclinations tilt to the NW and SE. From the samples obtained, some atomic absorption tests on metal cations were carried out, revealing interesting anomalies in the proportions of some priority elements, especially Pb (see Acevedo and Radoszta 1987). In the middle course of the Río Bolsa, on the right bank, immediately before receiving a tributary, there is an outcrop with abundant sulfides of a very pale yellow brass color, with large crystals with faces up to 3 mm that show the typical striations of pyrite, disseminated or concentrated, also associated with small quartz veins in very silicified sectors that cover the planes of weakness of black mudstone rocks with slaty cleavage. This sequence is affected by a smaller, neatly noticeable structure, which has a N-S course and a dip of 36° to the east. The pyrite is probably accompanied by chalcopyrite (although this was impossible to confirm due to the small size of the mineral grain), dispersed heterogeneously in the slaty mass. Pyrite is also associated with veins of hydrothermal quartz which, in turn, shows isomorphs of sulfur.

7.4.1.3   Bahía Valentín

In the southwest corner of Península Mitre, nestled between the Pirámide Hill to the west and the Montes Negros to the east, Bahía Valentín is located. On its wide beaches, the South America River and the Calavera Creek discharge.

Western coast of Bahía Valentín: The first rocky outcrops located next to the mouth of the Calavera creek—at the center west of the bay—are constituted by the aforementioned muddy fraction of the shales. They are black slates, pyritiferous. In this sector, the silicification of the rocks is conspicuous, traceable through quartz veins that reach 20 cm in thickness, conformable or not with the local structure.

Continuing toward the W and SW, important changes are highlighted. The structure is approximately NNE. Brecciation, fault mirrors with quartz veins, fractures in diverse scales, flexure cleavage, and other structures that infer a compression whose axis would be NE may be seen. The banded rocks—with local expression as slates—appear here very jointed and fissured, beginning of the large-scale mylonitization.

Further on from the previous site, on the cliffs, intense silicification is noticed. There are sectors where the banding of the layers is observed, with levels of calcarenites, pyrite bearing, in contact with a predominant fine muddy fraction—black silicified slabs—with some more disseminated pyrite. Feathery calcite veins cross these rocks. There are also numerous quartz veins, most of them conformable, that contain pyrite crystals up to 2 mm on each side. These veins are irregular in thickness, which varies between 0.5 and 15 cm. Other veins (somewhat corrugated) are made of pyrite, separated by sterile thicknesses of 5–25 cm.

The shales of the previous outcrops show abundant chlorite under the microscope and some sericite scales according to fine cleavage. The pyrite is fine but very abundant (15%), and scarce silty clasts correspond to quartz individuals with a section of 0.04 mm. Many folded quartz veins and silicification cores complete the petrographic observation. The classification corresponds to a carbonaceous argillaceous mudstones. To a lesser extent, calcarenites with quartz clasts (0.5 mm section) appear with angular shapes.

At the beach, the run of the pyritiferous layers along the coastline reaches a length of approximately 100 m, with thicknesses of 2 cm. The successive rocky outcrops banded from there show warps and kink-bands. Slaty schists and sandstones are encouraging for the observation of the structural attitude of cleavage and fracturing. The shales contain thin, siliceous balls, like phtanitic, with micaceous scales and pyrite crystals at the edges.

Eastern coast of Bahía Valentín: The first coastal outcrops, which immediately appear southeast of the mouth of the South America River, are composed of the lepto-metamorphic banded sequence, intruded by subvolcanic rhyodacitic porphyres being the strike contact ENE. Just above the contact zone between the two types of rock, there is abundant pyrite disseminated in the metasedimentary rocks, as well as pyrite microveins of thickness of the order of 0.5–1 mm, of a rather winding path but always accompanying the contact. The igneous rock shows a porphyritic texture with cataclastic quartz and ortho-phenocrysts, with corroded edges and filled with a vitreous to microcrystalline paste.

From the first porphyres to the south, light greenish-colored limestone schists appear strongly silicified and also chloritized, with jointing and mylonitization, having developed crenulation and transposition. The strike is N 70°E, and the dip is 15° to the southeast.

Outcrops similar to the previous ones, to the south of those, are also of calcareous schistose rocks, lazy very jointed. They exhibit a fracture cleavage that is tight, subvertical, bearing N 10°W. They have a slaty cleavage with strike of N 60°E (Fig. 7.4).

Silicification, oxidation, and chloritization of the rocks are largely developed in the outcrops of the region.

The eruptive mass, which is widely interdigitated with the metasediments of the country rocks, has also been transformed by intense mylonitization (Fig. 7.5).

The structural details indicated for the area, that is, slaty cleavage, crenulation, transposition, flexures, jointing, mylonitization and lava flow banding or fluidity of the eruptive rocks, show the same disposition, describing a single structure which is undoubtedly a reflection of a major cause or regional structure.

7.4.1.4 Synthesis of the Lithological and Structural Features of the Southern Coast of Península Mitre

From a sedimentary wedge of Tertiary rocks nestled in Bahía Sloggett to the east, the eastern continuation of the Deformed Complex of the Fuegian Andes appears represented in the region—between the Aguirre and Valentín bays—by two well-

Fig. 7.4 Eruptive rocks in advanced process of schistosity and jointing. Bahía Valentín. *Credit* R. D. Acevedo

Fig. 7.5 Crenulated eruptive rocks. Bahía Valentín. *Credit* R. D. Acevedo

defined lithostratigraphic entities: (a) a lepto-metamorphized sedimentary sequence and (b) subvolcanic acid eruptive rocks of porphyritic texture.

Regarding the first one, the sequence is composed of powerful sedimentary packages of stratified rocks. They consist of "flysch"-type turbiditic levels, in the same sense as Borrello (1969), composed of two clastic fractions, a coarser one, sandstones of the graywacke type, grayish-black to gray in color, and another one, very fine carbonaceous-colored black pelites, with very variable thicknesses and, sometimes, with calcareous interbedding.

The sedimentary sequence has been subject to a very low-grade regional metamorphism, visible under the microscope by the growth of tiny chlorite flakes in fine-grained rocks with clay cleavage. Eventually, some sericitic fibrils, most likely developed by thermal contact phenomena, may be detected.

The typical alternation between mudstones and fine sandstones is occasionally accompanied by carbonaceous and marly pellets, pillars of claystones and shales, fine sandstones, calcareous rocks, and rocks with siliceous nodules. Silicification phenomena that occur probably due to the remobilization of quartz during low-grade metamorphism are interspersed with phtanitic levels and ellipsoidal (deformed) chert nodules, normally conformable with respect to stratification, possibly as a replacement for calcareous forms.

Along the sequence sulfide oxidation phenomena have been observed, with the development of leaching zones, where pyrite is abundantly distributed and disseminated. These sulfides have been concentrated after the remobilization, caused both by the metamorphism and by the pressure and hydrothermalism linked to the porphyritic subvolcanic rocks. Quartz veins are added to these mineral appearances, conformable or unconformable, and the pyrite, that has recrystallized with greater abundance and size nearby such mineralization pathways, is related to the mentioned eruptive rocks.

The previous rocks, east of Bahía Aguirre, contain organic fragments that appear under the microscope to correspond to *Inoceramus* sp. Thus, those levels can be estimated to be of Cretaceous age, with caution (F. Medina, oral communication).

The regional structural guidelines have a general ENE direction. According to this dominant pattern, the courses of cleavage are aligned (slaty cleavage or axial plane cleavage) with inclination to the SSE (Fig. 7.6) of the contact between the eruptive bodies and metamorphic country rocks, of the angle of plunging, of a certain set of fractures, and even the coastal features, some streams and mountainous edges. The evident tectonic control extends to the mineralization of sulfides (pyrite), often linked to cleavage planes and veins and especially located in sectors where deformation and shearing folding are prominent features.

The style of folding is tight, in several orders. The folds are turned toward the NW quadrant and even lying down (Fig. 7.6), sometimes with sharp chevron-like hinges. Corrugation and kink-bands are common.

Occasionally, the deformation is so intense that it has produced mylonitization. The mylonites then occupy a position in accordance with the regional structural guidelines, and their appearance is linked to the fault and the contact zone between the eruptive entity and the lepto-metamorphic rocks, showing a considerable thickness of the order of 300 m in the crushing and mylonitization bands.

Considering the intense tectonic events that affected the Deformed Complex, obviously, fracturing is significant throughout the region. Its spatial attitude does not depart from the regional trend of the alignments; this is ENE, suspecting genetic linkage under the same process with events of intrusion, folding, metamorphism,

Fig. 7.6 Recumbent fold, 800 m ENE of the Bonpland River. Above: hinge surface; below: axial plane cleavage perpendicular to the primary structure in the hinge zone. *Credit* R. D. Acevedo

and fracturing (and subsequent reactivation of it) due to the parallelism shown by its respective attitudes.

7.4.1.5 Atocha, Campana, and Pirámide Hills

The "granitic porphyres": South of the Bolsa River, east of Bahía Aguirre, and the west sector of Bahía Valentín, a medium height but very steep mountain range extends in the direction E-W, composed of the Atocha, Campana, and Pirámide hills (Fig. 7.7).

Thick forests and mountain peatlands cover the metasedimentary sequence from the Bolsa River to the northern slopes of the mountain range. Some appearances of mudstones, sandstones, sandy limestones and marls, shales and schists have been

Fig. 7.7 Photograph of Mount Campana. *Credit* J. Azulay in Azulay and Azulay (2016)

detected until arriving at the proximities of a pinkish rock of subvolcanic eruptive type. In this sector, the contact between the two lithological entities is found.

Under the microscope, mudstones and sandstones (0.04 mm of average grain size) have (micro)clasts—mainly of quartz—with angular edges. An interesting amount of opaque minerals is also visible in the samples. The finer facies, of the clayey type, have developed a good cleavage—with iron hydroxide veins parallel to it—and the presence of chlorite and sericite.

Other samples are marly limestones in banks of a few cm thick, with 80% micritic carbonate and calcite sparite crystals. The remaining 20% is composed of quartz and feldspar clasts (plagioclase) of the order of 0.035 mm section.

There are also fine sandstones of the phtanitic type with oblong microstructures of 0.1×0.2 mm, filled by chert, dipped in a recrystallized (sericite) matrix and with fine cleavage.

Plentiful silica has been remobilized in the area. Numerous quartz veins cross the rocks in this site.

Toward the south of the previous point the eruptive body that comprises Mount Atocha appears. In a hand sample, it is noted that it is a rock of porphyritic structure, from pink to pinkish-gray color, with quartz phenocrysts and feldspar in felsic groundmass. Under the microscope albite and orthose appear, accompanying quartz, whose crystalline individuals are shown fractured and, commonly, with edges engulfed by the leucocratic groundmass. Chlorite and epidote have replaced a mineral that has probably been biotite. Opaque minerals (0.2 mm diameter), dispersed in the sericitized matrix, are also observed. The samples investigated correspond to a granitic porphyric alpha or beta (according to the IUGS classification) or rhyolitic–rhyodacitic. Frequently, these porphyres, already of an acid nature, can be found even more silicified, as it happens in the area of contact with their metasedimentary wall rocks. This contact, perfectly visible at the northern base of Mount

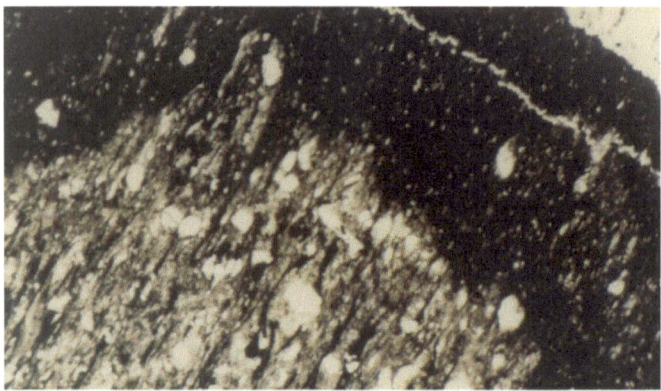

Fig. 7.8 Microphotography of a porphyre country rock contact (PPL, magnification ×100). *Credit* R. D. Acevedo

Atocha, is clear, observing true inclusions in the host rock that support the intrusive-ness of the eruptive body. Some of these xenoliths have been partially digested and strongly silicified. In the thin sections of the xenoliths, it can be seen that the crenu-lation is in a different position than in the box (which indicates that the xenoliths have been turned) but a posterior cleavage—oblique to the previous one—marked by the growth of sericite fibrils, which also modifies the porphyric, is common to the country rock porphyre–xenolith; this is an indicator that the last deformative event has been subsequent to the intrusion. Likewise, there are growths of prehnite in the breached zones of contact.

In addition to the aforementioned, there are—in the contact zone, inside the por-phyre—vitreous orbicules crushed and oriented according to the cleavage direction, which coincide with the axis of maximum elongation and deformation. Tiny crys-tals of feldspar and quartz have been formed inside the xenoliths by metasomatism. Quartz crystals have been microcrystallized, and a new generation of quartz coats the walls of the shear planes. In relation to the contact between xenoliths and the rock that contains them, the formation of chlorite, epidote, and sericite can be seen. The growth of this last mineral penetrates both rocks, forming cleavage planes. These planes cut angularly the surfaces of relictic corrugation of the xenoliths and are parallel to tiny oriented ellipsoidal orbicules that have been formed by the recrystallization of silica, disposed radially from a clastic core observed in the shales. It is emphasized that the elongation of the xenoliths (Fig. 7.8), the sericite guides that form a cleavage, and the elongation of the ellipsoidal silica micronodules in the country rocks keep a parallelism as well as the penetration of quartz and feldspar in the rock porphyritic in xenoliths. In the contact zone between the eruptive rock and the country rocks, shear and cataclasis are common and even mylonitization. This is especially observed at the E of Bahía Valentín and SW of Mount Atocha, on the slope of it toward the marine side. They are very fine granules, hard and coherent, flint like, and a whitish-gray color, with intense deformation of the primitive stratification planes.

The contact between the porphyre and the country rocks has, on Mount Atocha, the course of the great regional alignments, that is, ENE. Mount Pirámide is the easternmost of the three hills and the most accessible. Two fluvial streams flank this mountain toward the E and the W, the Luis and Beto creeks, respectively.

Blocks of rock and gravel constitute the top of the Pirámide Mount, in a very particular environment that represents a clear example of stone runs. In hand samples, it can be seen that they are yellowish-colored porphyres on weathered surface but more grayish in fresh fracture. The phenocrysts of quartz and feldspars are dipped in felsic cement. These porphyres contain differentiated greenish, elongated, small and with quartz crystallites included, which may well be xenoliths (true inclusions) witnesses of inhomogeneity of the paste.

Its oriented disposition marks an alignment by fluidity of the eruptive rock during its emplacement or through deformation.

Under the microscope, a porphyritic rock is observed, classified as a granitic porphyre on the basis of its siting and composition, constituted by 75% of phenocrysts and 25% of paste. The phenocrysts are mostly quartz (85%) with some lithic fragments (10%) and potassium feldspar (5%). The quartz crystals, of 0.5-mm mean section length, are moderately idiomorphic, with typical lightning extinction and even fractures. The lithic fragments, probably relicts of acid vulcanites with biotite, reach up to half a centimeter in diameter. The potassium feldspar, without twins (orthose), is frequently altered to clays. The felsic cement is in high proportion altered to a mineral with a flabelliform appearance (prehnite?).

The contact of the porphyres (broadly granitic) with their box is not observed as in the case of Mount Atocha; however, the break of the slope to the north of Mount Pirámide and the contiguous appearance of metasedimentary rocks suggests the eventual location of the covered contact.

The country rock is of the mudstone type, very dark gray to black. Its appearance is that of a hornfels—something banded—with injections of quartz lodged in conformable or unconformable veins. A thousand meters to the north, the mudstones become slates.

Toward the south, the eruptive rocks continue at least a thousand meters in isolation due to the peat bogs that cover the surface of the land.

To the west, Mount Campana seemed to keep more information regarding the intrusive relations between the porphyric rock and its hostees, pending such observations.

7.4.2 East Coast of the Península Mitre

(Montes Negros and western coast of the Straits of Le Maire, from Bahía Buen Suceso to Cabo San Diego and references to Isla de los Estados)

The east coast of Península Mitre to the easternmost end of Isla Grande is found between Montes Negros and Cabo San Diego, opposite to the Straits of Le Maire, whose margin is represented by the coasts of Isla de los Estados. Its expeditious geological description is also included in this chapter so as to enable the comparison between outcropping entities in one sector and another. More complete works on the Isla de los Estados geology can be found in Caminos (1975) and Caminos and Nullo (1979), and more recently, Ponce and Fernández (2013).

As in the southern coast, here the intrusive porphyric rocks and their lepto-metamorphic country rocks also appear, with the novelty of a third lithological type composed of fossiliferous sediments with inoceramids and ammonites of Late Cretaceous age.

(a) Schists, shales, metamudstones, and metasandstones: The oldest formational unit observed in the area is composed of sedimentary rocks, slightly metamorphosed but heavily folded. They have abundant quartz veins and veins in much greater quantity than the fossiliferous sedimentary levels of the northern area. Such rocks have been observed along the coast of Bahía Valentín and in the southern sector of the Montes Negros. They are powerful sedimentary packages deposited possibly in an environment corresponding to the edge of the continental shelf, genetically probable turbidites, composed of dark pelitic levels alternating with others of fine sandstone type, of light greenish-gray color. These rocks, as it has already been said, have been subject of a low-degree metamorphism and high deformation. It is still possible to distinguish the relictic banding of the stratified layers as well as to measure the angle formed by the axial plane slaty cleavage with the relict stratification.

In the outcrops observed in Bahía Valentín, quartz veins that cross the entire rocky area are common. These pyrite-bearing quartz veins frequently appear in disseminated form in the metasedimentary complex. It should be noted that the veins are more concentrated in the neighborhood of the porphyric bodies that lie in contact with the metamudstones and metasandstones psammites. It is also remarkable a high degree of general silicification of the sedimentary strata, which perhaps would be the product of the quartz remobilization during metamorphism.

Another outcrop assignable to this geological entity occurs north of the sandy Bahía Valentín. These are gray and black slates exposures and dark carbonaceous and calcareous pellets as well, with variations in greenish-gray tones. These layers are crossed by quartz veins deflected by the cleavage, which breaks in kink planes.

New signs of this lithological unit that corresponds to locally very folded rocks (Fig. 7.9) appear on the western slope of the edge to which the Muela hill belongs (which is part of the Montes Negros), in direct contact with porphyric rocks, as it will be seen below.

(b) Granitic, granodioritic, tonalitic (riodacitic and dacitic, sensu lato) porphyres: This subvolcanic unit is represented by rocks of acidic composition, porphyritic texture, and a partly intrusive emplacement. In hand samples, the pinkish to whitish color of a felsitic groundmass can be seen in which tiny crystals, mainly of quartz, are submerged. Under the microscope, phenocrysts of quartz, plagioclase, and potassium feldspar are distinguished in a leucocratic matrix. Quartz shows wavy extinction; their individuals are highly microcracked and

Fig. 7.9 Shales with higher
fissility and folding. *Credit*
R. D. Acevedo

Fig. 7.10 Views of the Montes Negros. *Credit* R. D. Acevedo

partially embayed by the groundmass. The plagioclase is andesine–oligoclase, proportionally predominant over alkaline feldspar (ortoclase). The matrix is of quartz–feldspar alkaline composition and has a marked alteration to chlorite, sericite, and epidote. These rocks can be classified as granitic–granodioritic or riodacitic porphyres. Compositionally, they grade into dacites.

These porphyric rocks constitute the main body of the southern sector of the Montes Negros (Fig. 7.10), and they are in intrusive contact with metasediments. There, the igneous mass penetrates, wedge-shaped, with strike N 40°E and dip 40°SE. The attitude of the contact surface is parallel to the schistosity of the country rock, which in turn is parallel to the regional alignment that follows a northeast direction.

The most accessible contact between the porphyres and the lepto-metamorphic country rock can be seen in the coastal outcrops located on the left bank of the South America River, at its mouth in Bahía Valentín. The intrusive character of the porphyre is substantial, with very visible features such as net contacts, xenoliths, and quartz injections. Although the contact surface is interfingered and folded, it has been possible to measure an approximate NE strike for it, more or less according to the regional structural parameters of folding, fracturing, and schistosity.

The eruptive rocks show intense deformation and folding of which the country rock of metamudstones and slates is obviously not exempt.

It is also necessary to highlight the presence of a mylonitized zone that involves both formations. Good profiles of the mylonites can be seen in the cliffs of the eastern coast of Bahía Valentín. The rock is of greenish color due to the alteration. Both in the original shale and in the deformed eruptive rock, there are different sets of cleavage—of crenulation and microjointing—maintaining the cleavage of the metamudstones and metasandstones with a strike between N 30°E and N 40°E, with a slight dip toward the SE.

In other outcrops, for example to the SE of the mountainous edge of the Muela hill, greenish-gray mudstones with rather more sandy facies appear, and carbonaceous transitions interspersed between the apophysis of the porphyric rocks. These sedimentary rocks are different from the metasedimentary rocks already described that do not show signs of metamorphism. The strike is N 40°E with a dip of 30° to the SE. They seem to be accommodated to the inflections of the eruptive tongues. The meaning of these layers is doubtful because there are no critical elements that allow to decide between the possibility that they are younger than the comparatively more metamorphosed rocks or that they are homologous layers that have been preserved from the accompanying deformation and metamorphism by the tempering effect that the eruptive rock masses have produced on such deformation. The contact relationships of these rocks with the mass of eruptive body only show apparently an overlap of the first one over the second, but the available information is not enough to make a decision.

(c) Fossils, sandstones, and fossiliferous calcarenites: Their outcrops extend from the central part of the Montes Negros to the coasts of Cabo San Diego and Bahía Tethis. In Montes Negros, these rocks comprise the mountainous mass, and further north to Bahía Tethis, the outcrops occur along the coasts and on the sides of inlets and bays. This powerful sedimentary sequence is locally folded upside down to the north, including also with flattened folds.

Concerning its lithology, this unit resembles the metasedimentary rocks located further south and described in point (a), but the structural character is certainly not the same, neither such rocks appear in contact with the porphyres. Its distinctive feature is the presence of bivalve fossils (Figs. 5.1, 5.4 and 5.5), and also some isolated ammonites as well, fairly well preserved, so that this entity would be comparable with what has been called the Beauvoir Formation (Camacho 1948). This definition was proposed for the northern area of the Beauvoir Range, and Furque (1966) later brought the concept of Beauvoir Formation to Bahía Tethis, applied to the so-called Estratos de Bahía Tethis ("Bahia Tethis Strata"; Furque and Camacho 1949). The specimens found in the southern coast of San Mauricio Cove and Montes Negros were given by the author to H. H. Camacho for their paleontology investigations. Professor Camacho determined the species Inoceramus steinmanni (Fig. 5.1) and *Inoceramus* sp. for successive fossiliferous localities, respectively, for which samples obtained in the southern coast of San Mauricio Cove can be considered as belonging to the Late Cretaceous (until the Campanian stage) according to verbal communication of the same author.

Fig. 7.11 Recumbent fold. *Credit* R. D. Acevedo

Detailed observations according to localities: The first outcrops from the south appear from a great regional alignment that, in ENE direction, continues from Bahía Valentín to the Patagones inlet. It is presumed that the fault contact between the sedimentary volcanic complex—described in points (a) and (b)—and the fossiliferous sedimentary rocks is defined by such an alignment. The layers of the lithological unit located to the north are very pale black shales, bearing N 45°E and dipping 32° to the SE, with phtanitic nodules and veins of milky quartz and rock crystal covering cavities. On the contrary, to the S of the alignment the layers are schistose banded, gray to green, of strike N 30°E and dip 35° to the SE, with structure of shear, kink-bands, cleavages of crenulation and fracture, and also with quartz injections.

Almost in the source of the South America River (in the Montes Negros), tube-shaped structures appear, approximately 5 mm in cross section, arranged concentrically inside a kind of cylinders of very dark carbonaceous calcareous mudstone, similar to the problematic ones found in Roncagli Point, Isla de los Estados. The first findings of these specimens date back to 1943 by Harrington in the northern shores of the island, in a geological environment which he called the "Serie Pizarreña".

Farther north, on the southern coast of Bahía Buen Suceso, gray, massive, intensely jointed mudstones and shales with pyrite appear in thin veins, with some relict features of primary sedimentation such as the characteristic granulometric banding, possibly containing some fossils (Fig. 5.2).

The structural style consists of soft folds. Locally, fracturing and refolding are very pronounced, and a mesofold lying with the axis plunging slightly to the south is observed (Fig. 7.11).

On the northern coast of Bahía Buen Suceso, the rocks are gray mudstones, approximately subhorizontal, interstratified with a slightly thicker facies of sandstones in smaller proportion, being able to observe the primary structure of cross

Fig. 7.12 Turbidites at San Mauricio Cove (clear layers have calcareous affinities and occasionally bearing undetermined ammonites). *Credit* J. Benito

stratification. Fracturing and shearing are common features. At the northern end of the bay, there are thick banks of gray sandy mudstones and thin black shale mantles.

To the south of the San Mauricio Cove, the sequence consists of fine, gray, lighter, or darker mudstones and sandstones interchangeably, with variable calcareous content and fossiliferous remains (Figs. 5.3 and 7.12), showing strong shear. Thin layers of fine sandstone of greenish-gray tones are interfingered with a silty–clayey facies. The rock contains some phtanitic nodules and calcareous lenses where bioturbation structures have been recognized.

In Herradura Cove, the intersection between the bedding planes (stratification) and cleavage determines acute angles. Calcareous levels are interspersed between the strata banded with mudstones and sandstones, with the latter being able to expose minor interstratal structures. Some sedimentary dikes cut across the gray mudstones.

In the route between San Mauricio Cove and Cabo San Diego, a monotonous folded and fractured succession of mudstone–sandstone layers emerges.

7.4.3 Northern Coast of Península Mitre

The Late Cretaceous of the Atlantic Ocean coast of the island of Tierra del Fuego is exposed along the marine shoreline from Península Mitre, following the NW direction until the Tertiary outcrops of the northeastern sector of the Isla Grande appear.

In the Bahía Tethis area, the "Estratos de Bahía Tethis" (informal denomination, Furque and Camacho 1949) would be the continuation of the fossiliferous entity that crops out between the Montes Negros and Cabo San Diego. These rocks could be consigned to the Beauvoir Formation, as proposed by Furque (1966) for Bahía Tethis.

Toward the WNW, the "Estratos de Policarpo" (informal denomination; also by Furque and Camacho 1949) occur. These authors established a Campanian–Danian age for this unit based on the study of its fossiliferous content).

7.4.3.1 Bahía Tethis

A thick bundle of dark carbonaceous mudstones ("Estratos de Bahía Tethis") appears strongly deformed and fractured, with subtle cleavage, affected by ultra-diagenesis processes and, probably, lepto-metamorphism. Some gray sandstones are interspersed, in blunt subordination to the finer, mantle-shaped, or nodular sedimentary rocks, also severely altered by the folding phenomena. The predominant strikes toward Cabo San Vicente are N 55° to N 65°E, with variable inclination according to the flanks of the folds (the most common one visible in the field is of mild 15°NW). Some samples seen in thin section reveal the presence of undetermined remains of foraminifera.

7.4.3.2 Puesto Tres Amigos

A basal conglomerate has been observed overlying the "Estratos de Bahía Tethis", composed of clasts of darker mudstone material, apparently coming from the finer sediments with cleavage that appear along the western coasts of the Straits of Le Maire or, locally, in Bahía Tethis. Upward in the sequence, new conglomeratic levels appear, interstratified with conglomerate sandstones and laminated sandstones, with the participation of finer, light greenish-brown, glauconitic conglomeratic sandstones (strikes of layers N 30/60°E and dip 20° to 25°NW). Here, the thick column that continues begins, at least, until Rancho Donata.

7.4.3.3 Caleta Falsa de Policarpo

At the foot of the Bilbao hill, the Caleta Falsa de Policarpo is located, a settlement area of Estancia Policarpo.

The rocks are represented there by a compact pack of fine marly, dark greenish-gray, glauconitic sandstones, with interbedded purer, 10-cm-thick, brown-colored sandy layers, carrying abundant dark gray calcareous nodules. In this sedimentary sequence, the pure sandy members represent only a small proportion of the total thickness. The nodules are very frequent. Its sizes, very diverse, range from 3 or 4 cm to half a meter in diameter. They are usually elongated according to the attitude of the stratification and lodged accordingly (Fig. 7.13).

The great uniformity of this sequence is noteworthy. Structurally, it consists of a homoclinal fold of little variable strike (from E-W to N 70°E) with dips that oscillate between 45° and 75° toward the NNW. The structure is strongly divided into many

Fig. 7.13   Calcareous sandy nodules. "Estratos de Policarpo." *Credit* R. D. Acevedo

places, according to the stratification, but in other places the strata are practically compact.

In the area where sea lions hunting took place, to the west of Caleta Falsa, there are thick layers of calcareous sandstones, with indeterminable remains of bivalve shells and very thin, dotted cylindrical structures, probably crinoids.

The strike of the layers is N 55°E and its dip of 45°NW. A marked alignment in the NNE direction of the mountain edges is truncated through mega-structures. The major fractures are oriented—from E to W—according to strikes of N 30°W, N 04°W, E-W, and N 80°W. At the western mouth of the cove, the sediments have an attitude of N 70°E (strike), 65°NW (dip). Marked joints of strikes N 08°E and N 10°E complete the observation.

On the eastern coasts of Caleta Falsa, the lithology is represented by greenish-gray calcareous sandstones of medium to fine grain size. The structure there is disturbed by a strong dislocation, marked by the follow-up of limestone members, carrying nodules, in very variable attitudes that, starting from a strike of N 45°E (with 20°NW of dip), reach N 70°E and dipping 33°–60° to southeast.

In the eastern tip of Caleta Falsa, the presence of indeterminate relicts of valves and organic structures such as those mentioned before is significant.

Toward the NE of the preceding site, always on the coast, another cape appears, composed of conglomeratic sandstones, with remains of fossil shells, of strike N 30°E and dip 58°NW, overlaid by sandstones, with fine conglomerates, of greenish-gray color, with glauconite, crossed by calcite veins.

Continuing toward the east, on the left margin of another unnamed stream, the same sandstones, with layers of nodules of limestone, whose follow-up is being carried out from Rancho Donata, are part of a folding mesostructure (axis N 75°E) that redefined the attitude of the subvertical layers toward the right margin of the stream that runs subsequently in favor of fracturing or faulting. The strike of the gray sandstone layers goes to N 65/70°E.

In the following units, the entity has a strike between N 60 and 70°E, with slightly variable dips up to vertical positions.

Folding structures appear toward the east (axes with a plunge of 15° and strike N 50°E). Then, although it has not been possible to establish the causes of the change, the layers have a strike E-W along a long stretch.

7.4.3.4 Between Estancia Policarpo and Policarpo River

To the east of Policarpo River, there are good rocky outcrops on the cliffs located on the right margin of its mouth in the sea.

Some lithological and structural observations of these exposures of the "Estratos de Policarpo" will be indicated below as locations A, B, B′, C, and D, easily located along a walking route.

Location A: thin to medium sandstones of dark greenish-gray color, of strike E-W and dip 60°S (outcrop thickness: 15 m), comparable to the outcropping rocks to the E of Caleta Falsa.

Location B: fine to medium granular limestone, of pinkish tones, folded.

Location B' (fossiliferous): light gray to gray limestones, strike N 70°E and dip 50°NW, upward in the sequence they become fossiliferous, being carriers of remains of "gryphaeas".

Location C: shelly sandstone bed in glauconitic sandstones, similar to the rocks of the sea lions hunting site, located to the W of Caleta Falsa.

Location D: green sandstones. At one end, erosion remnants appear as pedestals (as can be seen in Fig. 7.14).

The previous geological conditions continue without major structural deviations or facial variations until the mouth of an unnamed stream, dammed by the marine gravels. At this point, the calcareous glauconitic mudstones and sandstones with fossiliferous remains of Globigerina and bryozoans belonging to the "Estratos de Policarpo" (Fig. 7.15), similar in their dark green color to those of locality A, have a strike of N 30°E, dipping 30° toward the NW.

From there onward, the values of strike and dip of the strata are feature constant up to the outer western end of the Caleta Falsa de Policarpo, a verified feature through the observation of the alignments.

Crossing an extensive peat bog and moving away from the coast, the main building of Estancia Policarpo is found, located at the bottom of a closed cove.

7.4.3.5 Donata Hill

There are outcrops of thick calcareous sandstones, sabulitic to conglomeratic in their lower and upper levels, of whitish colors on weathered surfaces and of light brown

Fig. 7.14 Pedestal. 2 m of basal sandstones of attitude N 85°O and 42°N. Upward follows a paleosol of low thick. 2 m drift continues, covered by l, 5 m of peat and modern soil. *Credit* G. Giordanengo

to yellowish tones in fresh fracture. Clasts and matrix are predominantly calcareous, with some grains of feldspar and a characteristic that will be almost constant along the sequence—in spite of the facial variations: the presence of glauconite as it happens in the sandstones of the Claro River, in the Estancia Carmen area, north of the Beauvoir Range (Fig. 7.16).

The carefully observed profile has a thickness of 7.6 m, which is the maximum outcropping thickness. It begins with fine and medium conglomerate sandstones, graded, of 4 m thickness; 60 cm of thick sandstones follow, intensively welded, with remains of indeterminable shells and, finally, 3 m of conglomerate sandstones crossed by fine calcite veins (one set) and unconformable sandstones (two sets) of up to 1 cm in thickness, the unconformable ones being those of coarser size.

The general strike of the layers is N 85°W, and the dip of 15°/17° NNE. Upward, the beds have an attitude of N 75°W (strike) and 25°NE (dip).

The primary structures observed in this section consist of normal and graded stratification.

The last outcrops, toward the W of Donata hill, are very calcareous, and there are large folds in the wave cut platform (Fig. 7.17).

7.5 Isla de los Estados

The information presented in this chapter is intended to broaden the description of the author's direct observations in the eastern end of Península Mitre. For this reason, this chapter is purely complementary since there are previous studies and surveys carried out by Caminos and Nullo (1979) that contributed together with the studies

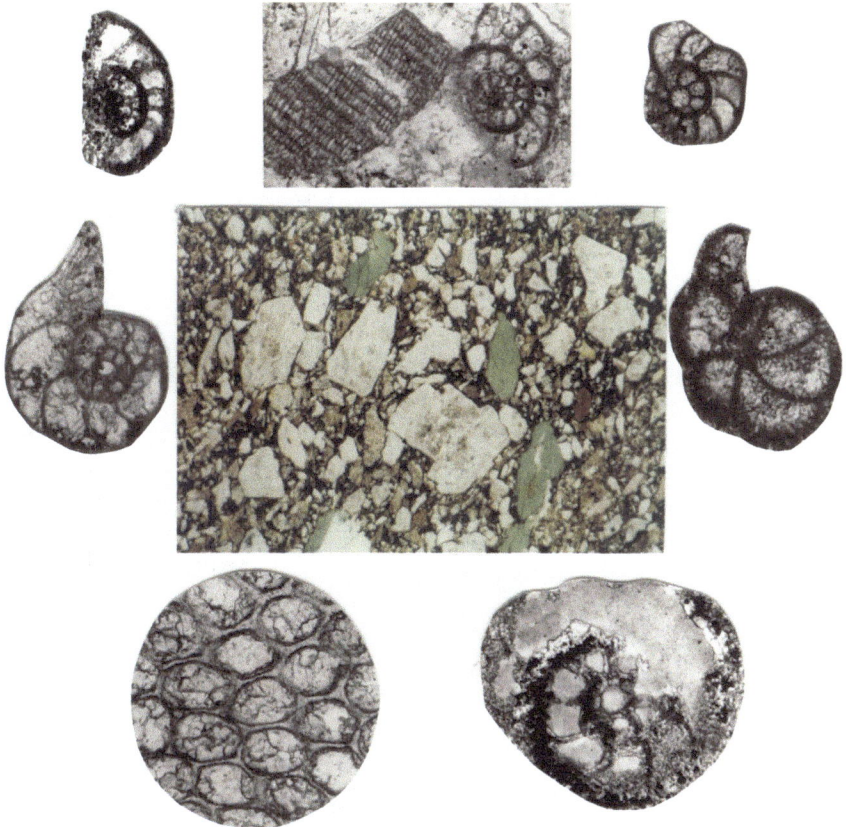

Fig. 7.15 Glauconitic sedimentary rocks bearing globigerina and bryozoans ("Estratos de Policarpo"). Magnifications: ×25 (glauconite), ×125 (foraminifera + bryozoans), ×125 (foraminifera), × 1500 bryozoans. *Credit* R. D. Acevedo

of Harrington (1943) to state the elements for the understanding of the geology of the island. In this way, Capitán Cánepa Cove, Celular Harbor, Bahía Vancouver, Cook Harbor, Hall Harbor, Cape Conway, Victorica Point, Roncagli Point, Bay Hoppner, and Cape San Antonio—Bahía Flinders—were recognized (Fig. 7.18).

The geology of the island is summarized as represented by two formational units, one of which, composed of eruptive rocks, occupies most of the territory, restricting the second, of sedimentary fossiliferous type, to the northern coastal rim. The entity mentioned in the foreground was correlated by Harrington (1943) with the "Serie Porfirítica", as described in Patagonia. Later, Caminos and Nullo (1979) included it into the Lemaire Formation (Borrello 1969, 1972), describing outcrops of "green-colored rocks and porphyric appearance, with folded mudstone interlayering." The other entity, formed by low-grade, metasedimentary rocks, of dark gray color, very fine grain and good cleavage, was identified by Harrington (1943) as "Serie Pizarreña"

Fig. 7.16 Glauconite
(green) in marine sandstones
from Estancia Carmen (Río
Claro). Microfossils 0.2 mm
diameter, PPL. *Credit* R. D.
Acevedo

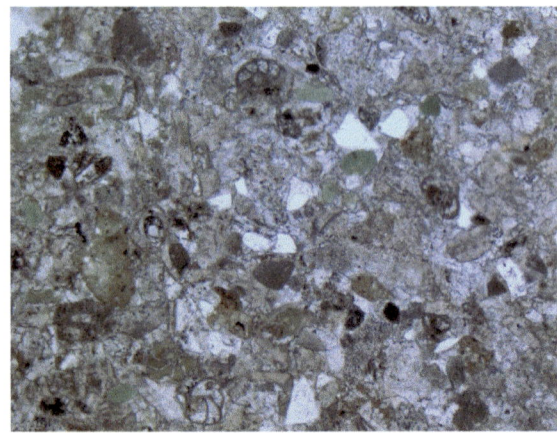

Fig. 7.17 "Estratos de Policarpo". Fossil calcareous facies with arches which include folding and
load marks in vertical strata. *Credit* A. P. Radoszta

and later assigned by Caminos and Nullo (1979) to the Beauvoir Formation (Borrello
1969, 1972), previously named as "Capas de Beauvoir" by Camacho (1948). The
following observations were performed in detail: (a) Lemaire Formation or "Serie
Porfirítica." In Celular Harbour, Bahía Vancouver, Cook Harbour, and Hoppner Harbour, the volcanosedimentary entity or Lemaire Formation was recognized. The
sampling was carried out on rocks of porphyritic structure, apple green to whitish
color, satin luster surface and very coherent. Under the microscope, a porphyric texture is observed. The quartz and plagioclase phenocrysts (oligoclase) are dipped in

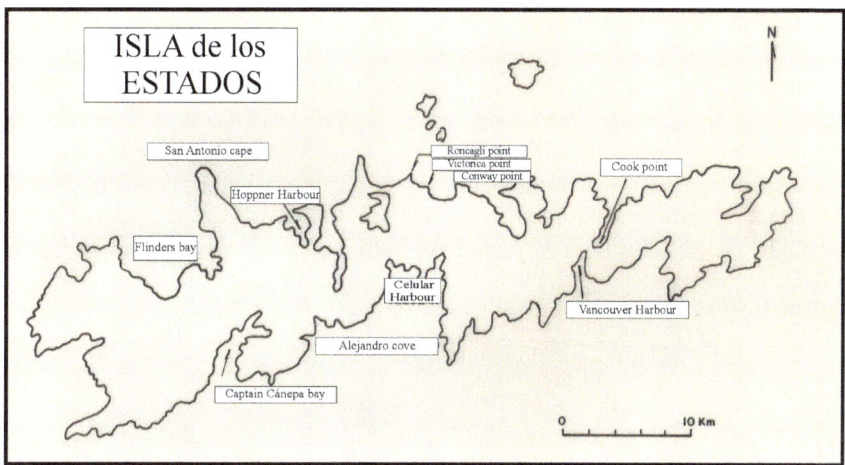

Fig. 7.18 Isla de los Estados with location of the visited localities

an acid groundmass. Growths of phyllosilicates (chlorite) have also been developed, oriented according to the superimposed secondary foliation.

Completing the composition of the rock, biotite, apatite, opaque minerals, titanite, epidote and calcite filling cavities. The size of the phenocrysts is variable, being very noticeable in its edges the corrosive action of the groundmass. Quartz, limpid and with secondary growth, commonly has wavy extinction. Plagioclases, fractured and altered, have the common polysynthetic mating. Occasionally, under the protection of the phenocrysts, it has recrystallized quartz together with sericite and chlorite flakes aligned under the effect of "pressure shadows." The crystals of quartz and plagioclase have been rotated. The chloritic–sericitic recrystallization obliterated the primary petrographic features. The microscopic observation made it possible to establish a clear difference in behavior between those rocks that absorbed the stress through the partial recrystallization of their pulp and the matrix, without disturbing the larger crystals, and a second type whose felsic base transmitted the effort to the phenocrysts that becomes deformed assimilating the cataclasis.

The samples obtained in Capitán Cánepa Cove, on the coasts of the northern lobe, 2000 m toward the SE of the summit of Mount Spegazzini, deserve a particular study. This sector has had to endure an intense tectonic activity, and large folds can be seen in the mountain escarpments. The mesoscopic structures (echelons and veins filled with quartz, rotated crystals) are related to shear movements and quartz crystallization under deforming conditions. Microscopically, a foliated texture is observed, in favor of the growth of stilpnomelane lamellae, accompanied by fibrils of sericite and chlorite in a felsic recrystallized silica groundmass. The thick facies consist of quartz crystals and plagioclase (acid andesine) deformed and with sutured edges. These individuals have rotated dragging with them a hair of straws of stilpnomelane that accompany the deformation. As accessory minerals, zircon and titanite have been identified. In a particular sample, a quartz vein of regular thickness was seen, with widening of it by crushing, and thinned core with small echelons that indicate the displacement. The schistosity there, marked by the stilpnomelane, crosses the quartz vein; that is, the growth of that was formed after the injection of quartz.

(b) Beauvoir Formation or "Serie Pizarreña": A second rocky type, circumscribed as it has already been said to the northern coasts of the island, could be examined with the coastal recognition of Flinders Bay (Cape San Antonio), Hoppner Harbour, points Roncagli, Victorica, and Conway and Cook Harbour. In these locations, outcrops of pellets of a dark gray color, folded crossed by guides of quartz and carbonate, were observed.

Samples of pelitic rocks were extracted with satin luster, which reveals an incipient metamorphism. Due to the small size of the clasts, its recognition under the microscope becomes uncertain. In spite of this, oblong structures (from 0.05 to 0.20 m) of quartz, reminiscent of the possible radiolarites found in Isla Grande, and plagioclase in small quantities, are recognized on the basis of extremely thin quartz.

Tiny sericite and chlorite fibrils define from their parallel arrangement surfaces of emergent schistosity, emphasized by the cleavage of clays. Sometimes cleavage is granulated until transposition, as in Conway Point, and it is not easy to define the degree of metamorphic reconstitution (Fig. 7.19). Accessory elements of these rocks are: apatite, opaque minerals—pyrite, goethite—and titanite. Pyrite is scarce, except in Flinders Bay where it has been observed disseminated in claystones and concentrated in nodules the size of small nuts. The veins of quartz and calcite are abundant. In Roncagli Point, calcite veins are posterior to the microjointing and previous to the deformation event that curved its cleavage and twinning. In this case, the calcite grew in clear association with silica removed by the slight metamorphism, having lodged calcite and quartz through thin fissures.

In Point Victorica, the quartz veins—1.5 cm thick—are also in precontact with the shear movement that gave rise to kink structures. Phyllosilicates accentuate the schistosity, cutting the veins.

Another noteworthy finding is the appearance of "problematics" in Roncagli Point, comparable to those mentioned for Montes Negros (Peninsula Mitre). These are remains of massive cylindrical elements, fine grain and dark gray color, elliptical base of 28 × 32 mm, with a tubular, eccentric duct, 2 × 3 mm in diameter, filled with calcite. The carrier mudstones of these problematics have an E-W strike and a N dip of 70°.

In the Hoppner and Cook harbours, mudstones folded in incongruity were observed, with sericite formation.

Conway Point: On the northern coast of Isla de los Estados, the slates found here have developed two surfaces "s" determined by crenulation cleavage and by the transposition of the same according to a reorientation of clayey elements and possible neomineralization of phyllosilicates.

A case of interest was observed in a tight anisopaque fold, of blunt hinges and long stretched flank that dips on its continuation in another short one. A contorted and disharmonic microfold of immediately smaller order is contained in the first one. These geometrical features have been controlled by the plastic behavior that facili-

Fig. 7.19 Crenulation of cleavage in black pelites of Roncagli and Conway points. (PPL, magnification ×100). *Credit* R. D. Acevedo

tates the flow of mutually sliding layers according to the fine strata in disharmonic microfolds.

The relationship with the larger structures of the place is not substantial. Even so, the interpretation can be advanced that the fold constitutes a trailing mesofolding where the inclination shown by the fold would give the sense of relative movement.

In essence, the bands of the rock sector that make up the fold are quartz and calcite, which, although undoubtedly recrystallized, suggest that they reflect the original composition.

An inequigranular structure has been developed, which is essentially granoblastic although granite relicts of quartz that show wavy extinction are preserved.

Therefore, the rock of the fold is a succession of alternating bands, respectively, of quartz pavements and in smaller quantity calcite, and vice versa, the average size of the quartz grain blasts, as well as those of calcite, is of the order of 0. 5 mm. Some larger quartz crystals are presumably relicts of the original grainy structure, with fractured edges.

Occasionally, quartz interpenetrates with calcite crystals.

It should finally be noted that there has been a post-crystalline deformation that is noticeable in the wavy extinction of quartz and in the curved cleavages of calcite.

In addition to the essential components that critically determine the aforementioned banding—this is calcite and quartz—the rock is characterized by a banding that is additive to the previous one (Fig. 7.20) that stands out remarkably and acts as a clear marker of the folded structure. These bands are composed of a grayish-brown amorphous material with a moderately high refractive index, superior to that of the balsam.

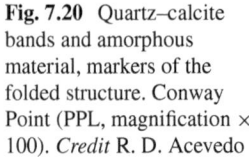

Fig. 7.20 Quartz–calcite bands and amorphous material, markers of the folded structure. Conway Point (PPL, magnification × 100). *Credit* R. D. Acevedo

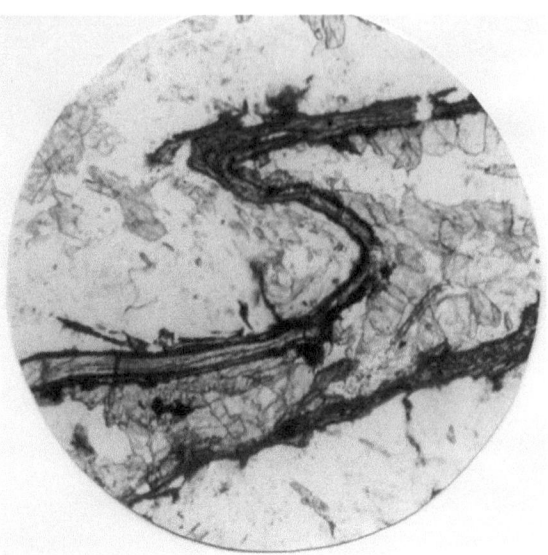

These bands are sometimes very clear, of constant thickness—0.05 mm—that sometimes become somewhat irregular.

Other mineral components of the rock are scarce. It is a case of fine crystals of actinolite of prismatic habit and green color. This amphibole variety composes fine bands interposed between the main rock banding. Actinolite needles penetrate to quartz. There are also epidote tablets, crystallized at the expense of the amorphous base, which may therefore contain calcium.

Consequently, the entity is defined in the following terms: rock folded in two orders, banded, with an alternate predominance of quartz and calcite recrystallized, the presence of quartz relicts and the existence of additional bands composed of amorphous, actinolite, and epidote material, respectively, being the first component the most significant and of greatest interest.

It is possible to test the history of the deformation and mineralization of the rock, with the idea that the amorphous material has been glass.

The shear stress controlled the fusion of the quartz–carbonate association. This has happened due to the frictional heat of the internal folding under low-grade metamorphic conditions, but certainly with an acute localization of heat and consequent high thermal state in the sliding surfaces. The glass formed, within this criterion, was placed in the contact between calcite and quartz. Subsequent cooling developed a devitrifying event that led to the formation of epidote and amphibole.

Two reactions would synthesize the case:

(a) Reaction stage between silica and calcium carbonate with iron and magnesium with glass formation without any traces of the possible formation of wollastonite being proved. The process has been a mutual reaction on the contact surface between quartz and calcite, elimination of carbon dioxide, all in a process of evident homology with (a) the formation of artificial glasses and (b) what happens in buchites produced in calcareous sandstones in contact with basaltic dykes.

(b) Stage of neocrystallization with formation of epidote and actinolite, which indicates a thermal low level related to the thermal stage of the metamorphism of this zone.

References

Acevedo RD, Radoszta AP (1987) Manifestaciones piríticas y anomalías metálicas en la costa sur de Tierra del Fuego. X° Congreso Geológico Argentino. Actas II: 227-230. San Carlos de Bariloche

Andersson JG (1906) Geological fragments from Tierra del Fuego. Bulletin Geological Institute University of Upsala, pp 169–83

Azulay J, Azulay J (2016) Tierra del Fuego. Península Mitre. Ed. Südpol, Museo Marítimo de Ushuaia, 120p

Bonarelli G (1917) Tierra del Fuego y sus turberas. Anales del Ministerio de Agricultura de la Nación. Sección Geología, Mineralogía y Minería, XII, N° 3 (Buenos Aires)

Borrello AV (1969) Los geosinclinales de la Argentina. Dirección Nacional de Geología y Minería, Anales 14 (Buenos Aires)

Borrello AV (1972) Cordillera Fueguina. En Geología Regional Argentina. Academia Nacional de Ciencias, pp 741–754 (Córdoba)

Camacho HH (1948) Geología del Lago Fagnano o Cami. Universidad Nacional de Buenos Aires. Biblioteca de la Facultad de Ciencias Exactas y Naturales. Ph. Thesis n° 543

Caminos R (1975) Tobas y pórfiros dinamometamorfizados de la Isla de los Estados, Tierra del Fuego. 6° Congreso Geológico Argentino 2:9–23 (Bahía Blanca)

Caminos R, Nullo F (1979) Descripción geológica de la Hoja 67e: Isla de los Estados. Servicio Geológico Nacional, Boletín n° 175 (Buenos Aires)

Caminos R, Haller M, Lapido O, Lizuain A, Page R, Ramos V (1981) Reconocimiento geológico de los Andes Fueguinos. Territorio Nacional de Tierra del Fuego. 8° Congreso Geológico Argentino, 3:754–786 (San Luis)

Furque G (1966) Algunos aspectos de la geología de Bahía Aguirre, Tierra del Fuego. Revista de la Asociación Geológica Argentina 21(1):61–66 (Buenos Aires)

Furque G, Camacho HH (1949) El Cretácico superior de la costa atlántica de Tierra del Fuego. Revista de la. Asociación Geológica Argentina III(4):263–395 (Buenos Aires)

Harrington HJ (1943) Observaciones geológicas en la Isla de los Estados. Museo Argentino de Ciencias Naturales. Anales, Tomo XLI. Geología. Publicación N° 29:29–52 (Buenos Aires)

Ponce F, Fernández M (2013) Climatic and environmental history of Isla de los Estados, Argentina. Springer Briefs in Earth System Sciences, South America and the Southern Hemisphere, 128p

Chapter 8
The Deformed Complex and the Formational Units

(a) The formational units that appear in the available geological literature, and that are found within the Deformed Complex of the Fuegian Andes, are the "Metamorfita Lapataia" (i.e., Lapataia Metamorphics), and the Alvear, Lemaire, Yahgan, and Beauvoir formations, "Estratos (strata) de Policarpo" and "Estratos de Bahía Tethis."

As an introduction to the elements of judgment that result from this study, the mentioned set of formational units is briefly reviewed below.

"Metamorfita Lapataia": These are strongly folded schistose rocks, injected by quartz. Located geographically in the bay of the same name in Canal Beagle, this unit extends from there toward the west, already outside the Argentine national territory and expanding in Chile.

It has been included by Kranck (1932a) in the Paleozoic "Metamorphic Schists." Borrello (1967, 1969, 1972) gave it the name with which it is currently known, suggesting a Jurassic age. Caminos et al. (1981) do not rule out a possible Paleozoic age for this unit.

Regardless of the age assigned, all authors agree that it is the oldest rocks at least in Argentine Tierra del Fuego.

Lemaire Formation: These are porphyritic eruptive rocks of acid composition with participation of sedimentary rocks already referred to in the works of Nordenskjöld (1905) and Quensel (1913), which correspond to the "quartziferous porphyres"—"Paleozoic" in the concept of Kranck (1932a)—, "Serie Porfirítica" (of the Late Jurassic of Isla de los Estados according to Harrington 1943) or "Porfírica," also called Tobífera Formation (of the Middle–Late Jurassic in Bruhn (1977), taken from Dalziel, according to the term coined by the Argentine oil geologists). Borrello (1967, 1969, 1972) called Lemaire Formation to this type of outcropping rocks in the Fuegian Andes, a name that Caminos (1980) used in Isla de los Estados. Furque (1967), on the other hand, called Lucio López Formation to the remnants of this entity on the southern coast of the Península Mitre.

© The Author(s), under exclusive licence to Springer Nature Switzerland AG 2019
R. D. Acevedo, *Geological Records of the Fuegian Andes Deformed Complex Framed in a Patagonian Orogenic Belt Regional Context*, SpringerBriefs in Earth System Sciences, https://doi.org/10.1007/978-3-030-00166-7_8

Alvear Formation: Camacho (1948) defined the Alvear Series when referring to the recognized sericitic schists in the Garibaldi and Spion Kop Mountain passes through the Cordillera Alvear-Lucas Bridges, assigning it a probable Jurassic age. Previously, Fester (1934) had assimilated the rocks of Cordillera Alvear to the Yahgan (Paleozoic) Formation of Kranck, accepting a possible extension of the deposit to Mesozoic times.

Yahgan Formation: This unit was founded by Kranck (1932a) who compares these layers (shales, graywackes, phyllitic slates, radiolarites, etc.) with the Monte Buckland Formation, also studied by Kranck in Chile and to which he assigned a Paleozoic age based upon stratigraphic observations (Kranck 1932b, c). Borrello (1967, 1968, 1972) named Monte Olivia Formation to the section more developed in a flysch facies of shales and graywackes. Dismissing the chronostratigraphic value of the foraminifera fossils that Kranck found in samples from the shores of the Murray Channel and that led the Finnish geologist to suggest a Paleozoic age for the Yahgan Formation as part of the "Central Cordillera schists." A later younger age was proposed by Katz and Watters (1966) who found in Navarino Island loose remains of *Inoceramus* sp. Halpern and Rex (1972) mentioned remains of Favrella americana or Favrella steinmanni on Gardiner Island (opposite to Picton Island). Caminos et al. (1981) divided this formation into lithofacies, and they mentioned *Chondrites* sp. observed in Ushuaia Península. According to the above, a Tithonian-Neocomian age could correspond to this entity. Di Benedetto (1973) extended the Yahgan Formation to the Beauvoir Mountains, beyond Lake Fagnano (Lake Cami).

Beauvoir Formation: This unit is known as "Flusch" of the "Marginal Cordillera" in the sense of Kranck (1932a) where it is located north of Lake Fagnano (Cami) and on the coasts of the Straits of Le Maire. The Beauvoir Series was named after Camacho (1948) who described blackish graywackes belonging to the early Senonian. Caminos and Nullo (1979) assigned to this formation the lands located to the north of Isla de los Estados. This unity contains rest of Belemnopsis patagoniensis (Late Jurassic–Early Cretaceous). The strata of Bahía Tethis (Furque and Camacho 1949) would also correspond to the Beauvoir Formation.

"Polycarp strata": These rocks are fine-grained sandstones up to conglomerates, fossiliferous, defined by Furque and Camacho (1949) to identify the Campanian sequence of the Atlantic coast of Tierra del Fuego.

(b) Critical aspects that allow a reconsideration of the Fuegian geology problem:

(1) In his purely geochronological work on samples from the SW of Isla Grande in Tierra del Fuego, Halpern (1973) published a date of 310 million years (Rb–Sr, total rock) for a metamorphic rock obtained at Bahía Pluschow. This information is very important and deserves to be highlighted because this is a very old age. Although it does not correspond to a rock of the Yahgan Formation because of the position in the Tierra del Fuego arch, a continuation of the rocks of the Yahgan Formation in the zone of Ushuaia is relevant to the dubious boundary between the rocks of Lapataia and the rocks of the Sierra de las Ovejas-Sierra de Martial, which correspond to the Yahgan Formation.

Such data leads to the possibility of considering, at least, the probable existence of an ancient metamorphic basement. The remaining results given by Halpern indicated more modern times, probably pointing to tectonic events and phases of magmatic activity. Halpern's radiometry data has been insufficiently evaluated by those who ventured a possible age of these marginal levels of the Deformed Complex.

(2)   The distinction made in the geological literature between the "Metamorfita Lapataia" and the Yahgan Formation is given by three characters that seem to be highlighted in the first entity. Greater metamorphism implied larger folding and more abundant presence of quartz. With regard to metamorphism, one of the distinctive features of the Lapataia schists is its greenish color given by intense chloritization of the rock. Some authors, citing the presence of biotite, argue the existence of a facies of greater regional metamorphism, without evaluating that their appearance (as in Río Olivia and Ushuaia peninsula) testifies to the metamorphic involvement by contact of an intrusive body although the type of bedrock that gave way to the schists has not been evaluated. In relation to the supposed greater deformation, the observations have been made along the coasts, where the marine abrasion enhances the folding effect. The rocks belonging to the Lemaire Formation or the Yahgan Formation, according to the structural position where they are found, also show a marked folding. The remaining criterion, that is to say, the abundant participation of quartz veins in the "Metamorfita Lapataia," is a detail that does not exclude the entity from the east that, by the way, is not devoid of the same quartz. In addition, the relatively close intrusiveness of the lithodemic unit of Cordillera Darwin batholith could also have influenced this high proportion. The quartz present both in the area of Lapataia and in the area of the Yahgan Formation, such as the obvious case of the Río Pipo, can have two origins: Either it has been provided or it has been segregated during the folding as it happens in certain gneisses of higher metamorphic degree in other regions. Actually, both things happened; perhaps the quartz of the discordant veins was received from the batholith or simply remobilized.

(3)   Specifying the limits of the Yahgan Formation is, without doubts, a delicate issue. Perhaps this is the reason why there are so many interpretations concerning his age.

Dana (1848) described a fossil in Nassau Bay (in the Cape Horn region) which was later assigned an Aptian–Albian age.

Kranck (1932a) preferred to compare the Yahgan Formation with the Mount Buckland Formation, to which it had assigned Paleozoic age based on the findings of foraminifera (between Navarino and Hoste islands) whose biochron, it was later proved, had not chronostratigraphic values. This impairment and the discovery of new paleontological material, especially from Bahía Tekenika, influenced those who, later, in the absence of fossil record in Argentine territory, resorted to those to correlate the sterile sequence under study. For this reason, the generalized opinion of considering the whole sequence mapped in the original Yahgan Formation of Kranck as deposited between the Late Jurassic and the Early Cretaceous lacks good support.

Although already Borrello (1969, 1972) separated the outcroppings of the surroundings of Ushuaia—which he called the Monte Olivia Formation—from the rest, this decision was based for its distinction upon the characters of the Fuegian flysch, a criterion collected by Caminos et al. (1981) in the lithofacial division of the whole group. It would also be prudent, in the general conception of a regional metasedimentary complex, to bring together all those sectors in which no determinative fossils of defined age have appeared, whose limits would coincide—at least as far as Pampa de los Indios, the eastern mouth of Canal Beagle—with Kranck's concept of the Yahgan Formation, which he himself had founded.

Toward the south, those limits would not, for now, go far beyond Canal Beagle, on the margin of the islands located to the SE of the Moat Channel. To the north, beyond the great line of Carbajal-Tierra Mayor-Lashifashaj and its possible continuation in the valley of the Moat River, the rocks may very well be a continuation of the Yahgan Formation, but there are critical elements that establish differences (Biel et al. 2007) against the single morphological–structural question of the great valley.

The concept that the Sierra de Sorondo is composed of a formation and that the Cordillera Alvear-Lucas Bridges it is so by another has not only resulted from the general strike of the Carbajal-Tierra Mayor-Lashifashaj valley, almost certainly bounded by tectonic features, but also by the presence of rocks of volcanic origin in the second unit (Caminos et al. 1981). However, most of these leucobasalts in the broad sense are intrusive and are also found at the northern edge of the Sierra de Sorondo (Mount Olivia) and the Alvear-Lucas Bridges mountain range (see Quartino et al. 1989). The location of these eruptive rocks is surely tectonic in relation to the fault lines or structures of the NNE section of the Olivia River, subparallel to Lake Escondido and transversal to them in the Carbajal-Tierra Mayor-Lashifashaj valley.

(4) As for the continuation to the E of the Yahgan Formation, when Kranck (1932a) extended it to Península Mitre, he did not notice that it had a younger age there than it had been presumed for that entity. Kranck had not been in neither Península Mitre before, nor in the Alvear mountain range.

Old findings of the present author of macroscopic remains of "*Inoceramus* sp." on the eastern coast of Bahía Aguirre suggest that it is possible to assume a Jurassic–Cretaceous age for these levels of fine mudstones and sandstones. Although no more data from the Lucio López Range is available than the information obtained by Hurtado et al. (5th PEOAF Territorial Museum of Ushuaia)—that allow to suppose the continuity between two entities that do not differ much in their lithology—it is probable that the alignment of the López River valley (or the Moat valley) separates two lithofacial units, the western one (Formation Yahgan) and another one, eastern and younger. Unfortunately, only one tectonic wedge with Tertiary layers (Andersson 1906; Caminos et al. 1981) appears in Bahía Sloggett, where the López River discharges.

(5) It is desirable to highlight the intrusive nature (at least verified in some sectors) of the porphyry band of rhyodacite composition that forms the Atocha, Campana, and Pirámide hills, extending to the Montes Negros, to the east, and

the western mouth of Bahía Aguirre. These acidic eruptive rocks are hosted by mudstone–sandstone metasedimentary rocks, the whole being strongly folded. They also contain packages of non-metamorphosed sediments. Kranck (1932a) who had studied "quartziferous porphyries" in an environment of mylonites (on the northern slope of Mount Olivia) linked them to those of Península Mitre in his map, and, since then, this correlation became widespread.

The truth is that, as already stated above, near the basic intrusive body of Mount Olivia, Kranck described a rock that was apparently an acid porphyry. When sampled during the present study such an outcrop proved what Kranck himself had said, this is that supposed phenocrysts of albite without any deformation lie in a highly deformed mass which suggests that it is a metasomatic origin of these crystals, or that a subsequent thermal metamorphism caused such homogenization in true phenocrysts. Therefore, the geographical location of this igneous manifestation should be restricted to Península Mitre and Isla de los Estados. Regarding its geological location, it is possible that there have been pulses of acid eruptivity with underwater sedimentation cycles which in this case suggests that there may have been lavas or domes that summarize intrusive and effusive characters.

(6) The observations made in Montes Negros have revealed data of fundamental significance to clarify some points concerning the geology of the eastern end of the Fuegian archipelago. On the one hand, the discovery of fossils ("*Inoceramus steinmanni*") in the central and northern summits of this mountainous edge extends the deposits of the Late Cretaceous much more to the south than was supposed before. On the other hand, this sequence (which does not differ with respect to the metasedimentites of the south of Península Mitre more than in its structural style of less tight folds and less presence of quartz in veins) is not altered by the previous granitic intrusion. An alignment that propagates in direction ENE between Bahía Valentín and the Patagones inlet separates these fossiliferous sedimentites (to the north) from the volcanosedimentary complex to the south.

One unknown fact that must be cleared is the circumstance that the fossiliferous entity of the Late Cretaceous is apparently in agreement with the general inclination to the south under the set of porphyries and sedimentites that are considered as older units, perhaps of Tithonian-Neocomian age. The solution to the problem would be in the presence of an anticline of magnitude, overturned to the north, in a way that the outcrops that are found to the south are of older age.

At last, reference should be made here to the geology of South Georgia Island (Thomson et al. 1982) where the Annenkov Island Formation has been presented. On the basis of its faunal content, this entity is placed above the Cumberland Bay Formation, assuming an Early Cretaceous age for the whole and establishing its correlation with the Yahgan Formation. Among the fossils described, specimens of "*Inoceramus* sp." stand out. These specimens are identical to those collected in the Montes Negros, and thus, a comparison of the Annenkov Formation with the Montes Negros strata would be appropriate (perhaps the basal section of the Beauvoir Formation outcropping to the N), instead of such one with the Yahgan Formation.

8.1   The Regional Metamorphism

Katz and Watters (1966) in Isla Navarino and Caminos et al. (1981) in Isla Grande defined a regional metamorphism of prehnite–pumpellyite degree for the Yahgan Formation.

It is noted, in addition to the above, the almost constant presence of chlorite throughout the microscopic inspection of metasedimentary rock samples from Tierra del Fuego. Chlorite, regardless of its appearance as an alteration mineral, defines the range of regional metamorphism. The local presence of other minerals—such as sericite (muscovite), biotite, garnet and andalusite—has other connotations, which will be analyzed in detail in the next chapter.

This question is fundamental as the proven thermal metamorphism places critically the issue of the definition in the degree of regional metamorphism.

References

Andersson JG (1906) Geological fragments from Tierra del Fuego. Bulletin Geological Institute University of Upsala, pp 169–83

Biel C, Subías I, Fanlo I, Acevedo RD (2007) Mineralogical characterization of Lemaire and Yahgán Formations, Tierra del Fuego, vol Libro de Resúmenes 23. Argentina GeoSur, Santiago de Chile

Borrello AV (1967) Estado actual del conocimiento geológico del flysch en la Argentina. Revista del Museo de La Plata. Tomo VI. Geología, N° 44: 125–153. La Plata

Borrello AV (1969) Los geosinclinales de la Argentina. Dirección Nacional de Geología y Minería, Anales 14. Buenos Aires

Borrello AV (1972) Cordillera Fueguina. En Geología Regional Argentina. Academia Nacional de Ciencias, Córdoba, pp 741–754

Camacho HH (1948) Geología del Lago Fagnano o Cami. Universidad Nacional de Buenos Aires. Biblioteca de la Facultad de Ciencias Exactas y Naturales. Ph.D. Thesis n° 543

Caminos R (1980) Cordillera Fueguina. En Geología Regional Argentina. Academia Nacional de Ciencias. Córdoba, II: 1463–1501

Caminos R, Nullo F (1979) Descripción geológica de la Hoja 67e: Isla de los Estados. Servicio Geológico Nacional, Boletín n° 175. Buenos Aires

Caminos R, Haller M, Lapido O, Lizuain A, Page R, Ramos V (1981) Reconocimiento geológico de los Andes Fueguinos. Territorio Nacional de Tierra del Fuego. 8° Congreso Geológico Argentino, 3: 754–786. San Luis

Di Benedetto HJ (1973) Mapa geológico de la Cuenca Austral. Yacimientos Petrolíferos Fiscales, Buenos Aires

Fester GA (1934) La Cordillera Alvear y el valle de Tierra Mayor. Revista Minera, VI: 49–64. Buenos Aires

Furque G, Camacho HH (1949) El Cretácico superior de la costa atlántica de Tierra del Fuego. Revista de la. Asociación Geológica Argentina, III(4): 263–395. Buenos Aires

Halpern M, Rex DC (1972) Time of folding of the Yahgan Formation and age of the Tekenika Beds, southern Chile, South America. Geol Soc Am Bull 83:1881–1886

Harrington HJ (1943) Observaciones geológicas en la Isla de los Estados. Museo Argentino de Ciencias Naturales. Anales, Tomo XLI. Geología. Publicación N° 29: 29–52. Buenos Aires

Katz HR, Watters WA (1966) Geological investigation of the Yahgan Formation (Upper Mesozoic) and associated igneous rocks of Navarino Island, Southern Chile. NZ J Geol Geophys 9(3):323–359

Kranck EH (1932a) Geological investigations in the Cordillera of Tierra del Fuego. Acta Geogr
    4(2):231p; Helsinki
Kranck EH (1932b) Sur quelques roches a radiolaires de la Terre de Feu. Extrait du Bulletin de la
    Societe geologique de France, 5ᵉ serie, 2: 275–283
Kranck EH (1932c) Sur quelques roches à Radiolaires de la Terre de Feu. Geologisch-
    paläontologisches Institut der Universität Basel. Societe Geologique de France, pp 275–283
Nordenskjöld O (1905) Die krystallinen gesteine der Magellanslander. Wiss Ergebn Exp Magell
    1(6):175240
Quartino BJ, Acevedo RD, Scalabrini Ortiz J (1989) Rocas eruptivas volcanógenas entre Monte
    Olivia y Paso Garibaldi, Isla Grande de Tierra del Fuego. Rev de la Asoc Geol Argent 44(3–4):328-
    335. Buenos Aires
Quensel P (1913) Die Quarzporphyr- und Porphyroidformation in Südpatagonien und Feuerland.
    Bull Geol Inst Upps 12:9–40
Thomson MRA, Tanner PWG, Rex DC (1982) Fossil and radiometric evidence for ages of deposition
    and metamorphism of sedimentary sequences on South Georgia. In: Craddock C (ed) Antarctic
    geoscience. University of Wisconsin Press, Madison, pp 177–184

Chapter 9
The Magmatic Rocks Probably Corresponding to the Andean Batholith and the Associated Metamorphic Contact Phenomena

The knowledge about the presence of outcropping plutonic rocks in the Argentine side of the Fuegian Andes is limited exclusively to the references about the dioritic bodies, in a wider sense, of Mount Jeu-Jepén (immediately to the ESE of Lake Fagnano), Spion Kop (in the Lucas Bridges Range, on Paso Harberton) hills, and Estancia Túnel, to the E of Ushuaia. The first two were initially described by Camacho (1948), a study in which they returned (Caminos et al. 1981). Regarding the plutonic rocks of Estancia Túnel, Morelli and Azcuy mentioned (see Orquera et al. 1977) "a black plutonic rock of the gabbro type" outcropping in a place called Lancha Packewaia. This intrusive body, studied by Acevedo et al. (1989), was attributed to the Andean Batholith, and it would be comparable with those of the Santa Rosa Plutonic Complex, located in front of Ushuaia, Navarino Island, and described by Suárez et al. (1982, 1985), although new evidence in González-Guillot et al. (2018) would support other interpretations.

This chapter refers to the issues related to local plutonism nearby Ushuaia, mentioning the critical localities in the finding of key elements in the petrographic and structural investigation of the intrusiveness and its thermal effects on the Deformed Complex which served as country rock.

9.1 The Plutonic Rocks

The intrusive bodies of two localities, respectively, Estancia Túnel, to the E of the mouth of the Olivia River, and the other one exposed by the quarrying and preparation of work in the Ushuaia Peninsula, will be described below. Everything expressed means that the igneous activity immediately to the N of Canal Beagle is important and very significant because it is outside the plutonic axis located further south.

© The Author(s), under exclusive licence to Springer Nature Switzerland AG 2019
R. D. Acevedo, *Geological Records of the Fuegian Andes Deformed Complex Framed in a Patagonian Orogenic Belt Regional Context*, SpringerBriefs in Earth System Sciences, https://doi.org/10.1007/978-3-030-00166-7_9

9.1.1 Estancia Túnel

At the coasts of Canal Beagle—to the W of the Escarpados beacon and up to the vicinity of Estancia Túnel—an ultramafic body crops out, crossed by clear veins, which intruded a sedimentary lepto-metamorphized set, with a predominance of mudstone and graywacke facies, in a typical marine sequence, in that the microscopic presence of biotite and garnet was verified, an issue that will be discussed later.

It is evident that the plutonic rocks and their veins belong to a body of which there are elements that indicate greater dimension in depth. The best exhibitions of the rocks are found at the cliffs of the coast.

The most striking of the outcrops is the relatively homogeneous continuity, except grain size difference, of a rock mainly composed of hornblende in crystals up to 7 cm in length and, more commonly, 3–5 cm. In these cases of such a thick grain variety, no leucocratic mineral is observed in the field. In a first field impression is a hornblendite with a coarse texture lacking internal orientation, to the point that the thick granular interlocking, to the naked eye, shows that some of the amphibole crystals interfinger. Varieties finer than the one described were observed more toward the west, near the Escarpados beacon, which under the microscope showed to be similar to the rock in question.

The dominant color of the type rock is dark green to blackish, sometimes with brown tones evidently due to weathering. The described continuity is interrupted by the very abundant white leucocratic veins, very contrasting, and by less abundant spots or veins of yellowish green color, composed of epidote, from 0.5 to 1 cm thick. Some net surfaces of joints in the terrain are determined by the planar disposition of the epidote.

The clear veins are, most of the time, planar. In other cases, they are curved or crossed or anastomosed in dominant subvertical positions or with strong inclination. The thickness is variable. In the field, they have all the appearance of late veins of the same forming process of the ultramafic plutonic rocks (Fig. 9.1).

Excluding the epidote concentrations, at first sight the lithological types are three: (a) the ultramafic rock, (b) the relatively fine grain of the white veins, and (c) a variety that corresponds to the sum of the other two types with a medium color index, which is usually in continuity with the type first in the vicinity of white veins.

The ultramafic plutonic rock is characterized by quite pronounced variations in the proportions of the minerals and in the presence of granules, noticeable in different thin sections. This is partly due to the presence of very coarse grain size with medium grain size rocks.

The two main minerals are hornblende and clinopyroxene, in such proportions that in most cases the first predominates over the second, to the point that in some of the thin sections no pyroxene is observed. Except for samples that exhibit a significant proportion of leucocratic minerals, the ultramafic plutonic rocks are defined as a pyroxene hornblendite with transitions to hornblendite–pyroxenite.

Fig. 9.1 Block of ultramafic rock with syenitic veins. Estancia Túnel. *Credit* C. D. Coto

The synthesis of the mineralogical composition is the following: hornblende, pyroxene, magnetite, pyrite, apatite, titanite, biotite–phlogopite, albite, orthose, and epidote.

The structure is between panalotriomorphic and hypidiomorphic, with a relation between pyroxene and hornblende of not very clear definition, since gradual phenomena of the type of uralitization are not observed. They are only contacts between grains without evident order of crystallization, or pseudo-inclusions of hornblende of transitional contours, existing optical continuity between the different pseudo-inclusions which indicate the possibility of a pyroxene replacement by amphibole. Amphibole and clinopyroxene vary from anhedral to euhedral, the latter being smaller.

Clinopyroxene is normal, colorless, positive biaxial with a gamma angle C never found above 44°, which may mean a presumably augite character. Fine Schiller structure is seldom found.

The hornblende has a more common pleochroism z: green and greenish yellow, x: green; also observed—in some cases—carrying slightly bluish tones. A fibrous amphibole strongly colored in green and blue tones has been observed (z: light greenish to green, x: yellowish greenish pale) that crosses common hornblende grains, and even to the interstitial crystallization epidote. Opaque minerals are magnetite and pyrite also, less abundant, partly replaced by iron oxide. Its distribution is irregular; they are xenomorphic showing that they corrode the mafic minerals. Its size is variable (from microscopic to 3 mm), always lower than that of pyroxene or amphibole. Although it is not a constant rule, there is a high frequency of association between the opaque minerals and the apatite crystals (Fig. 9.2). This apatite is fresh or with a phenomenon of opacity and alteration of brown color that inhibits any figure of interference. Apatite crystals are euhedral to subhedral.

Fig. 9.2 Photomicrograph
of hornblendite with opaque
minerals and apatite. (PPL,
magnification × 100). *Credit*
R. D. Acevedo

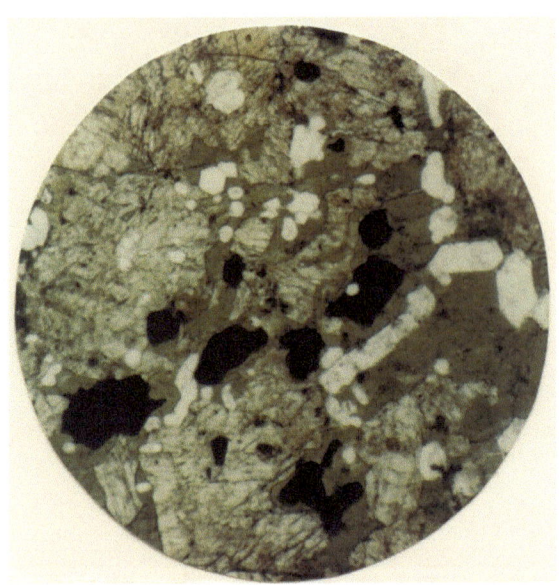

The biotite, of very low pleochroism and of chestnut color (better defined as biotite–phlogopite), is scarce, standing out that it has a smaller size than the crystals of pyroxene and amphibole, but forming part of the primary texture of the rock.

In many cuts, no leucocratic minerals are observed but, sometimes, as trapped in the main granular structure of the rock, nests or interstices with alkali, sodium, and potassium feldspars are found.

This presence of feldspars, due to their proportions in the type ultramafic rock, does not affect the proposed classification of hornblendite/pyroxenite.

The alteration, in addition to what has been mentioned about the apatite, is reduced to the formation of epidote at the expense of the mafic minerals and, sometimes, in veins that cut even the opaque minerals. It should be noted that the epidote is found in two forms: one in the alteration of mafic minerals or in veins, and another as a continuous optics mineral that is part of the granular aggregate of the rock, possibly by filling intergranular spaces left by the building crystalline amphibole and less pyroxene. This epidote is crossed by the aforementioned fibrous amphibole.

The most opposite rock from the compositional point of view with respect to the one described here is that one which constitutes the white veins. The rock is of subalkaline syenite composition because the dominant minerals (on average by 95%) are albite and orthose varyingly perthitic; these two minerals being in an estimated proportion of 60–40% in the order expressed. The concept of subalkalinity refers to the lack of plagioclase with higher calcium content than albite. Erratic presence of idiomorphic amphibole, dispersed opaque minerals, and small idiomorphic rhombuses of titanite are added to these minerals.

The structure of the syenitic vein results from the existence of large euhedral to subhedral albite and euhedral perthitic orthose, among which there is a microgranular base with the same mineralogical species, but with an obvious predominance of albite; in the case of clearly differentiated veins and late magmatic episodes, it matters the presence of quartz. Up to now known veins lack quartz and also clinopyroxene. As a magmatic late mineral, or alteration, there is epidote.

The third lithological type is intermediate between the ultramafic hornblendite–pyroxenite and the rock of syenitic composition of the veins, which, as it will be seen later, is interesting for the petrological interpretation.

This third type is characterized because the melanocratic part, which usually exceeds 70–80%, lacks pyroxene, being composed of hornblende and magnetite–pyrite. The leucocratic part is composed of albite in greater quantity than orthose, sometimes perthitic, which has normal contacts with the plagioclase or corrodes it. The alteration is of epidote and, in cases of very marked processes, in the formation of sericite at the expense of the leucocratic aggregate.

The deformation is not significant in the rocks described. Already in field observations, it has been seen that the veins do not show structures with deformation subsequent to them. This does not mean that cataclastic features are observed under the microscope. This is how they are seen as irregular flexures in the amphibole of the hornblendite and within a common non-cataclastic scenario. There are resultant features in the syenites of the veins such as flame-shaped twinning in plagioclase, wavy extinctions and perthitization, perhaps increased by the stimulus of the deformation. Contrary to what Suárez et al. (1982) have stated for the Santa Rosa Plutonic Complex in Navarino Island (just south of Ushuaia), the deformation of the body is interpreted as "auto-cataclasis," produced by the emplacement of the magma. The field data reinforce this conception since the deformation at the edges of the body is there more heightened.

The leucocratic veins that cut the ultramafic body have the character of dioritic or syenitic porphyres in general, with albite and with the characteristic of the predominance of sodium over potassium according to the modal composition.

Recently obtained samples from an outcropping plutonic rock between Escarpados beacon and Point Jones, carrying a coarse texture, belong to a rock composed of quartz, plagioclase (oligoclase), and some potassium feldspar (orthose), with hornblende and biotite. As an accessory mineral, there is apatite. Alterations to chlorite, epidote, and clays are common. The rock can be classified as a quartz diorite.

Table 9.1 shows chemical analysis data from samples of ultramafic rocks (analysis 1 and 2), their veins (analyzes 4 and 5), and an intermediate variety (analysis 3).

With regard to the metallic content of the ultramafic rocks, it was interesting to obtain some information related to chromium and nickel. Atomic absorption spectrophotometry was used to determine 199, 166, and 154 ppm of chromium and 30, 58, and 102 ppm of nickel for hornblendites from Estancia Túnel. Being the samples of surface rocks, the data is interesting, especially since it encourages improvement for depth, especially because the possible olivine level is not found in the exposed levels.

Table 9.1 Chemical analysis of the intrusive rocks of the Estancia Túnel

	1	2	3	4	5
SiO_2	38.29	41.56	42.03	60.60	64.01
Al_2O_3	14.81	11.78	18.44	18.67	18.40
CaO	10.61	12.26	11.44	3.56	1.30
MgO	12.12	11.91	5.20	1.26	1.05
Fe_2O_3	6.92	7.05	6.39	2.02	1.64
FeO	7.16	7.01	6.50	0.96	0.77
Na_2O	1.64	1.20	2.34	6.25	7.62
K_2O	1.26	1.26	1.44	4.08	3.38
TiO_2	1.54	1.46	1.18	0.30	0.22
MnO	0.20	0.16	0.22	0.05	0.05
P_2O_5	0.42	0.22	0.66	0.10	0.04
H_2O^+	0.99	0.71	0.52	0.47	0.34
H_2O^-	0.04	0.12	0.10	0.07	0.05
Fe_2O_3T	14.87	14.84	13.61	3.09	2.38

9.1.2   Península Ushuaia

As it was previously seen, small intrusive bodies have been uncovered by cleaning works in the peninsula located in front of the city of Ushuaia. Samples of these rocks have been prepared for their microscopic observation, having basically observed two petrographic types:

9.1.2.1   Hornblendes and Related Varieties

These rocks have a texture ranging from coarse to porphyric. The proportion of amphibole varies widely up to 85% of its total composition; it occurs in phenocrysts of tabular habit or as elongated festoons, often observing pseudomorphs of a lumpy mineral that possibly corresponds to the first generation of amphiboles. Leucocratic minerals are also invariably present in the samples investigated. Quartz, interstitial, is sometimes accompanied by potassium feldspar and plagioclase (oligoclase–andesine). As accessory components, apatite, titanite, and opaque minerals are present. The most common alterations are kaolinization, chloritization, sericitization, and formation of epidote, titanite, and a lumpy mineral unidentifiable by petrographic methods. In the porphyritic variety, the texture of the paste, constituted mainly by plagioclase tablets, is pilotaxitic.

9.1.2.2 Dioritic Porphyres

They are formed in roughly equivalent parts by phenocrysts and groundmass. The phenocrysts are plagioclase euhedral to subhedral, very altered, of andesine composition, with occasional albite or Carlsbad twins, possibly of a zonal nature; and scarce hornblende. Regarding the paste, it is basically quartz, with some twinned, acidic plagioclase. Accessory components are apatite, zircon, and opaque minerals. As secondary minerals, chlorite, epidote, and calcite were found.

9.2 The Phenomena of Thermal Metamorphism

9.2.1 Estancia Túnel

The rocks that are to the east and to the west of the outcropping places of the ultramafic intrusion and that, in first instance, can be taken as country rocks of the same one are a succession of black to gray mudstones and sandstones, with thicknesses from a few centimeters to half a meter in a folded structure. In the field and the observation of hand samples, they have no striking characters with respect to a simple interpretation as sedimentary rocks. It escapes to this the observation of cleavage at an angle with the stratification, the appreciation of a slight schistose texture, and, for the case of some localities, the strong impression of hornfels as it is very well observed at the Estancia Túnel. Rocks are strongly welded there. It turns out then that there are two structural and mineralogenic processes for microscopic observation in detail. The first is defined by low-grade shales in conjunction with folding and the second by the recrystallization attributed to the intrusion. These two processes are sometimes not clearly distinguishable, whereas in other cases the difference is clear due to the presence of minerals that can be identified as indicators of contact metamorphism such as garnet, biotite, well-recrystallized muscovite, and, in some cases, epidote. The second phenomenon of crystallization results in the formation of biotite and garnet as more critical minerals (Fig. 9.3) and of muscovite and epidote as accompanying minerals.

Biotite forms isolated pleochroic lamellae in only one approximation to the salt-and-pepper-texture, or its laminae are oriented according to the schistosity by mimetic crystallization. With the well-recrystallized muscovite, the latter mainly occurs. The epidote is very fine-grained and variable in quantity, possibly because of the original composition of the rock. As for garnet, it is very small grain-sized, microscopic, yellowish, often with central interior turbidity, possibly by idiomorphic iron oxide in its vast majority. It is dispersed or concentrated in contacts between thicknesses of different granule sizes. There is also a third and curious form of location, which is the attachment to threads of opaque minerals, presumably magnetite. This last location and the yellowish color suggest the presence of iron in the composition of the garnet, without having other data.

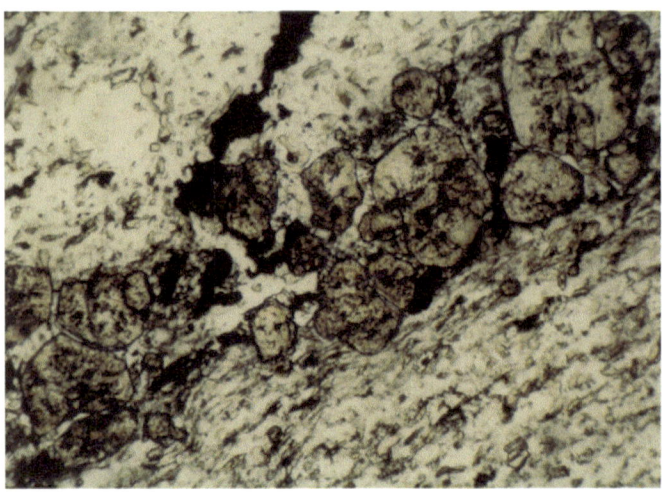

Fig. 9.3 Garnet and biotite hornfels. Olivia River, PPL, magnification × 375 *Credit* R. D. Acevedo

It has been observed that in some cases the garnet, when growing in the pelitic aggregate, has curved the schistosity with good suggestion that it is a late or post-kinetic crystallization.

These schists with plentiful garnet crystals and sometimes with biotite—not so abundant—were originally found at the mouth of the Olivia River. For this reason, they raised the question of whether it was the detection of a regional metamorphism of greater degree than the one known in the area. The aforementioned petrographic characters of the disposition of the biotite and the discovery of these minerals in the rocks with evident aspect of hornfels from Estancia Túnel turned the interpretation toward the understanding of contact metamorphism, which has also been detected halfway between Estancia Túnel and Encajonado River. In this way, no evidence is provided that would justify the suspicion of regional metamorphism of a higher grade than that which could be defined as sericite–chlorite.

The interpretation made about contact metamorphism has the character of the detection of a localized thermal episode, evidently post-tectonic with respect to the acute deformation of the shales. This is in agreement with the appreciation of the minor deformation, although not totally absent, of the ultramafic intrusion and its associated rocks.

Contact relations between the ultrabasic plutonic rocks and their metasedimentary country rock may be seen in the field. The hornblendite–pyroxenite body of Túnel (Lancha Packewaia) shows an intrusive contact with the rocks where it is lodged. The contact area has been revised on the coasts of Canal Beagle, highlighting the following characteristics:

(a) The eastern contact on the beach to the west of the building of Estancia Túnel is hidden; the contact between the plutonic rocks and their hornfels country rock cannot be seen. The deformation there is of a high nature due to the intrusion of the ultramafic body, which at its mineral composition adds at these levels a higher concentration of pyrite and oxidized copper, located the latter in joining surfaces. Under the microscope, a biotitic hornfels can be observed, with a granular texture (quartz) to salt-and-pepper-texture (quartz and biotite) in whose composition some individuals of plagioclase (andesine) and apatite and opaques as accessory minerals also participate. Another contact metamorphic mineral, besides the biotite, is sericite, and as an alteration, abundant chlorite and epidote.

(b) Western contact. Halfway between the Escarpados beacon and Point Jones, a wide stream marks the contact between the plutonic rocks and their country rock. This contact is visible (even on the slopes of the right bank of the creek) and neat, clear, with some interfingering of the intrusive and regional rock, banded to finely laminated, of strike N 30°–50° W and variable dip of up to 75° SW. Under the microscope, garnetiferous hornfels with a granular texture are observed, with a higher proportion of recrystallized quartz and opaque minerals as iron oxides among them. As alteration minerals, chlorites and epidote may be cited. As a product of metamorphism, there are sericite and garnet, not having been recognized in the sample biotite specimens. Quartz and zoisite veins and alkaline feldspar without twins cross the rocks. As structural data, corrugation cleavage is observed.

9.2.2 Península Ushuaia

Regarding the rocks of the intrusion country rock, in the Ushuaia Peninsula, it is interesting to study the contact metamorphic progression that, starting at subtle levels of chlorite, reaches the formation of muscovite, biotite, and garnet.

The original sedimentary rocks have been with high probability siltstones and quartzitic sandstones of orthoquartzitic type (some possibly radiolaritic) and arkoses. No relics of lithic fragments of vulcanites were observed although high recrystallization may have masked the possibility of retaining the relict of acid volcanic rocks.

The size, shape, and arrangement of the crystals of these rocks are represented by the typical salt-and-pepper-texture of the biotite hornfels.

Opaque minerals (mostly pyrite) in fine grains are abundant. The presence of epidote as alteration mineral, like chlorite, is also common.

The appearance of garnet is linked to the existence of calcite and its location, bordering on levels of phtanite, interpreted by many as ancient radiolarites.

At last, a quick look at the intrusion and its country rock allows to visualize the forced diapiric character of the hornblenditic body, which has arched in an anticline and metamorphized to the layers where it was emplaced.

9.2.3 Lake Acigami (Lake Roca, Tierra Del Fuego National Park)

Kranck (1932) studied the field relationships between the metamorphic basement, the quartziferous porphyres, and the Yahgan Formation near the Argentine border, something more toward the NW, in Chilean territory, citing the presence of biotite as a determinative mineral of the regional metamorphism that modified the sequence. Previously, Lovisato (1883) and Nordenskjöld (1905) had mentioned the identification of some individuals of andalusite in this region.

The objective of the treatment of this problem has been to look for possible indications of distinctive metamorphism and folding of a different geological entity ("Metamorfita Lapataia") with respect to the typical features of the Yahgan and Lemaire Formations.

Starting from the southeast end of Lake Acigami, the northeast coasts of this lake were visited in the field, until arriving at Hito XXIV ("hitos" are milestones marking the international border), and from there, by the international boundary until the first mountainous edge that borders the lake.

A sample (corresponding to a shale) extracted from a point near Hito XXIV, observed in petrographic section, would reveal with reserves the existence of andalusite (Olivero et al. 1997), a mineral lodged in vast quantities among the schistose bands of amphibole, chlorite and muscovite, quartz, and plagioclase. The mineral in question is altered from the center of the crystal toward its periphery to another fine and lumpy mineral. As accessories, some iron oxides are found, being also epidote identified as an alteration mineral.

About the occurrence of andalusite in these shales, its metamorphic origin is highly possible through a thermal contact effect on a plutonic rock which does not outcrop here or it has not been discovered yet.

References

Acevedo RD, Quartino G, Coto C (1989) La intrusión ultramáfica de Estancia Túnel y el significado de la presencia de biotita y granate en la Isla Grande de Tierra del Fuego. Acta Geológica Lilloana 17(1):21–36. San Miguel del Tucumán

Camacho HH (1948) Geología del Lago Fagnano o Cami. Ph. Thesis no. 543, Universidad Nacional de Buenos Aires, Biblioteca de la Facultad de Ciencias Exactas y Naturales

Caminos R, Haller M, Lapido O, Lizuain A, Page R, Ramos V (1981) Reconocimiento geológico de los Andes Fueguinos. Territorio Nacional de Tierra del Fuego. VIII Congreso Geológico Argentino, vol 3. San Luis, pp 754–786

González Guillot M, Ghiglione M, Escayola M, Martins Pimentel M, Mortensen J, Acevedo RD (2018) Ushuaia Pluton: Magma diversification, emplacement and relation with regional tectonics in the southernmost Andes. J S Am Earth Sci (submitted)

Kranck EH (1932) Geological investigations in the Cordillera of Tierra del Fuego. Acta Geogr 4(2):231 Helsinki

Lovisato D (1883) Apuntes geológicos sobre la Isla de los Estados. In: Bove G (ed) Expedición Austral Argentina, Instituto Geográfico Argentino, Buenos Aires, pp 47–53

Nordenskjöld O (1905) Die krystallinen gesteine der Magellanslander. Wiss Ergebn Exp Magell 1(6):175240

Olivero EB, Acevedo RD, Martinioni DR (1997) Geología y estructura del Mesozoico de Bahía Ensenada, Tierra del Fuego. Revista de la Asociación Geológica Argentina 52(2):169–179

Chapter 10
Nature and Synthesis of the Results Obtained

The preceding chapters have tried to present the author's direct observation data, locality by locality, about the set of eruptive and metamorphic rocks in the Fuegian Andes. This has been done in this way based upon the idea that an integration of the data as well as interpretations and emerging ideas can be, to an acceptable extent, separated from the data themselves. Starting from the gathering of bibliographic information layers gives a guideline of the previous investigations. It was also intended that in the progress of the studies the concepts of those previous elements left a margin of freedom for the search of new critical data even in well-known sites. It is the sense of having adopted from the beginning the concept of Deformed Complex of the Fuegian Andes as a comprehensive denomination of the lithological group that extends from west to east from Bahía Lapataia to the Straits of Le Maire, to which some direct observations in Isla de los Estados were added, with only a complementary character.

Working conditions in the studied area can be understood as very special given the need to add multiple focused discussion from the territorial point of view while harmonizing the whole idea. It is worth remembering that the discovery of fossils of the Late Cretaceous in Montes Negros, that made clearer all attempts of correlation, was made late in the development of the studies and that the knowledge of contact metamorphism was also addressed when the finding of the first indicator minerals of this phenomenon had been preceded by the mere verification in numerous places of low-grade regional metamorphism concomitant with intense deformation and folding.

For the scope of the results obtained, the examination system of successive localities has the value, in the opinion of the author, of the contribution of reliable data. It also stands out in this scope to have recognized for the first time areas of the Fuegian Andes, such as the case of Península Mitre, of very difficult access, which was the result of the development of expeditions in the 1980s, with pedestrian tours of the extreme projects east of Península Mitre named PEOAF 2, 3, and 4, thanks to the organization of the Museum of Ushuaia. Other new findings were not the result of approaching remote and isolated areas but the identification of rocks in very acces-

© The Author(s), under exclusive licence to Springer Nature Switzerland AG 2019   95
R. D. Acevedo, *Geological Records of the Fuegian Andes Deformed Complex Framed in a Patagonian Orogenic Belt Regional Context*, SpringerBriefs in Earth System Sciences, https://doi.org/10.1007/978-3-030-00166-7_10

Fig. 10.1   Mount Vinciguerra, 1470 m.s.n.m. *Credit* Wikipedia

sible places by quarrying or removal of the Quaternary cover, such as the case of the Península Ushuaia where the igneous rocks and the contact aureole were clearly visible.

It remains to be added, in order to clarify the scope of this work, that observations were made in the Vinciguerra mountains (Fig. 10.1) and the ranges of the same name from Lapataia to Ushuaia, in the Sierra de Sorondo, in the cross section of the Cordillera Alvear-Lucas Bridges, in the cuts that the National Route number 3 offers, as well in Estancia Harberton and the No-Top Range to Pampa de Los Indios, Bahía Sloggett, and from Bahía Aguirre to Policarpo Cove, passing through the Montes Negros. As it is noted, the mountain ranges of Noguera, Irigoyen, and Lucio López do not count on any observation in this work (see González-Guillot et al. 2009). In the same way, the Beauvoir Range was left aside (see Martinioni et al. 1999). From this brief elucidation, it is evident that the geological knowledge of the Fuegian Andes is a long-term undertaking of which the present study is only a minor part.

10.1   The Deformed Complex and Its Subdivision

A summary of the results achieved and the proposed linkages is as follows:

The Deformed Complex of the Fuegian Andes includes the lands located from the coasts of Canal Beagle and the Mar Argentino (Argentine Sea) from the south (excluding the Tertiary layers with coal mantles or cyclothems from Bahía Sloggett) to the southern edge of Lake Fagnano and its continuation toward the East. The powerful fossiliferous Late Cretaceous sequence although folded, is preferred in this work to be disregarded for the moment from this complex, which, toward the west, reaches the international boundary, and there are no doubts of its continuity farther in that direction.

The criterion reached in this work has been to subdivide the Deformed Complex into two sectors: western and eastern, separated tentatively by the Moat and López rivers. This last region, between the valleys of these two rivers, was only briefly visited due to operational reasons; therefore, there is no evidence for a definitive delimitation between the two sectors of the complex. The lepto-metamorphized sediments and the structural style are similar in both, there being a very definite, distinctive factor that consists in the presence of abundant acid porphyritic rocks in the eastern sector.

The western sector of the Deformed Complex comprises the Cordillera Alvear-Lucas Bridges; the mountain ranges of Valdivieso, Vinciguerra and Sorondo, Martial, Andorra, de las Ovejas, del Toro and No-Top ranges; isolated hills such as the Susana, Tonelli, and Cuchilla mountains; and the respective piedmont sectors. Lepto-metapelites and lepto-metapsamites are the most representative rocks of the region. Apparently, there would be not enough reasons to define the formational foundation within the sector although mapping units based on metamorphism could be distinguished as well as the presence of pre-dynamic eruptive rocks of mesosilicic and basic composition. Although this last criterion loses value, the valley of Lashifashaj and the area of Rancho Hambre-Lake Escondido responds to locations bounded by tectonic features in lines or bands of weakness or fracturing corresponding to the two alignments that coincide with the two geographical places mentioned.

The FADC-ES includes, at least, the cliffs of Bahía Aguirre, Atocha, Campana, and Pirámide hills and part of the Montes Negros, continuing in the central and southern sectors of Isla de los Estados. The set of porphyritic rocks of acidic composition is in clear relation of intrusiveness with low-grade metasedimentary rocks.

In summary, the Fuegian Andes, leaving aside the mountain ranges that were not studied in this work, comprise of two large units: (a) The Deformed Complex subdivided into the eastern sector and the western sector, both without fossil elements, with the exception of microscopic fossil elements found in Bahía Aguirre, that is, in the eastern sector. It is clarified that the western sector includes the area of Lapataia whose formational individuality with respect to the Yahgan Formation, which is the rest of the western sector, is doubtful; and (b) the mountain sector between the middle latitude of Montes Negros and Policarpo creek, also deformed, with fossils indicative of Late Cretaceous age.

10.2 Lithology: Stratified Layers

The western sector of the Deformed Complex of the Fuegian Andes is mainly composed of mudstones and sandstones with an E-W general strike that has developed a slaty cleavage in favor of a regional tectonic event with overturned folds, in a broad sense, toward the north.

It is important to note the occurrence of calcareous rocks as sedimentary interbedded strata due to deformation in the area of the dam of the Olivia River and in the Martial Mountains (see Chap. 3) or as segregations in veins produced by the incipient remobilization by metamorphism. These rocks have quartz veins concordant with the folding—the oldest—and discordant, subsequently. It is proposed for the former an origin by exudation or segregation during the folding and lepto-metamorphism and for the later a relocation of the previous material.

As for the FADC-ES, the rocks outcropping between Bahía Aguirre and the southern part of the Montes Negros are also stratified mudstones and sandstones, somewhat carbonaceous and highly silicified nearby the eruptive outcrops. The development of cleavage is conspicuous. The marly and calcareous layers are also frequent, especially in the Bahía Aguirre area.

At last, in the northern part of Montes Negros, a thick sequence of mudstones and sandstones (also fractured and folded) with significant participation of calcareous layers of reduced thickness, ammonite-bearing, emerges without solution of continuity. In the area of Policarpo, the limestones are more frequent and thicker, with abundant fauna of gryphaea, foraminifera, and bryozoans. These layers (assigned to the Late Cretaceous) that are found from the northern half of the Montes Negros to the Policarpo area are considered, as it has already been said, outside the set called Complex Deformed, which poses greater questions due to the absence of fossils so far (except the microscopic remains found east of Bahía Aguirre).

10.3 Lithology: The Pre-tectonic Eruptive Rocks

In the western sector of the Deformed Complex (FADC-WS), between Mount Olivia and Lake Escondido emerges a succession of eruptive bodies of basic to mesosilicic composition, pre-tectonic with respect to the main deformation event of the region that produced the axial plane cleavage of the larger folds. They can be classified as leucobasalts to melandesites, with textures ranging from officinal and subophitic to pilotaxitic, intergranular, and intersertal. These bodies are emplaced in country rocks of metasedimentary rocks (mudstones and sandstones) of low metamorphic grade, comprising lense shapes of massive appearance in its cores to semi-schist and greenschists toward the edges.

In the case of the body located at the base of Mount Olivia, its subvolcanic character is noted, with inclusions of the country rock (as xenoliths) in the eruptive mass. However, even in the case of localized intrusions in areas of tectonic weakness, their position in the chronostratigraphic column is still being discussed, since on one hand they are shown as younger entities than the regional metasedimentary rocks, and on the other hand, they are considered in alternation with sedimentation and belonging to a bimodal volcanism.

As for the pre-tectonic eruptive rocks that are located in the FADC-ES, it is relevant that, in strong contrast to the previous ones, they are acid porphyres (of granite composition, sensu lato) whose outcrops extend in an E-W direction, coinciding with the great regional alignments, along more than 40 km, on the southern coast of Península Mitre, intruding—at least in part—the lepto-metamorphic rocks that serve as country rock or interfingering with them. The occurrence of this great eruptive appearance is interpreted as domes that in their final stage have expanded.

A question still to be solved is the temporal location of the eruption activity in both sectors of the Deformed Complex that adds to another problem—not even solved—that is the age of the rocks in which the pre-tectonic igneous bodies are emplaced. This is concomitant with the inclusion of the Yahgan Formation in the Western Deformed Complex with the doubt of adding or not to the whole of the Cordillera Alvear-Lucas Bridges. The rocks of Lapataia Bay could also be included in the above, at least until the eastern limit of them is reliably confirmed (Acevedo 1988). The rocks of the FADC-ES coincide with the formational concept of the Lemaire Formation with the addition of the high participation of sedimentary rocks linked to the porphyres.

10.4 The Post-tectonic Eruptive Rocks Related to the Andean Batholith

Plutonic bodies of reduced size and coarse texture emerge on the coasts of Canal Beagle. In the area of Estancia Túnel, these intrusive bodies are hornblendite–pyroxenites crossed by veins of a differentiated magma of syenitic composition. Some 12 km to the west, in the Ushuaia Peninsula, small bodies of hornblendites and quartz diorites with porphyritic variations appear. This post-tectonic intrusive activity, together with the plutonic rocks of Jeu Jepén and Spion Kop, located farther north, is outcrops quite far from the Andean Batholith, whose plutonic axis lies farther south, in the area of Cape Horn.

One of the fundamental characteristics of these bodies is the production of contact metamorphism verified in the area of Estancia Túnel and near Ushuaia.

10.5 Contact Metamorphism

The appearance of biotite, garnet (and probably andalusite) in low-grade metasedimentary rocks in the area near Ushuaia has been analyzed as a marker of a second metamorphic event, of a thermal type, in rocks sometimes cropping out (such as in the case of the plutonic rocks of Túnel-Packewaia or the granular processes of the Península Ushuaia) or hidden from observation (linked to the possible presence of andalusite to the NNE of Lake Acigami). The importance of the corroboration of contact metamorphism extends to considerations about general metamorphism and in the sense that it excludes biotite as an indicator of a greater regional metamorphism.

10.6   Structure

10.6.1   Major Structures

The main features of the major structure are the following:

(a) Alignment of mountainous ranges. In the FADC-ES, the parallelism of the mountainous ranges that make up the Fuegian Andes is notorious. To the south of Lake Fagnano, the ranges of Valdivieso–Alvear-Lucas Bridges and Vinciguerra–Sorondo are successively aligned, separated by the great depression of the Carbajal-Tierra Mayor-Lashifashaj valley. In the eastern sector (Península Mitre), the eruptive outcrops—which form the highest mountains—mark the orientation of the larger structure, with strikes predominant in an ENE direction.

(b) Depressed belts. Three large regional depressions, also extending in the direction of approximately E-W orientation, are occupied, respectively, from north to south by Lake Fagnano, the Carbajal-Tierra Mayor-Lashifashaj valley and Canal Beagle.

As for the Carbajal-Tierra Mayor-Lashifashaj valley, it is also bounded by the higher altitudes of the Fuegian Andes, as it has already been mentioned. At last, Canal Beagle up to its eastern mouth in front of Pampa de los Indios also occupies an extensive stripe between the Isla Grande to the north and Isla Navarino to the south, also showing a mountainous relief.

The physiographic observation makes it possible to infer the non-exclusive possibility that these are tectonic horst or grabens housed between elevated tectonic blocks represented, respectively, by the depressed belts and the Andean Ranges. The traditionally accepted model consists of a strike-slip fault. This idea does not agree with the outcrops of the Santa Rosa pluton (on Navarino Island) and the small bodies in the Península Ushuaia, as opposed to the previous plutonic rock, which would almost prove that there was no displacement after the intrusions or suggested that these intrusions are not related (González-Guillot 2018). The well-known concept of strike-slip belts in the depressed areas does not oppose the idea of the uplifting blocks because they may have been successive episodes, the second of them with inclined planes toward the south.

(c) Folding structure. Folds of regional expression, visible in the minor and intermediate orders or supposed in major orders, characterize the Deformed Complex of the Fuegian Andes.

The case of Mount Olivia anticlinorium synthesizes the folding of the western sector of the complex. The folds are asymmetrical, with the northern limb abrupt, steep, or well overturned, with axial planes of general strike E-W and south dip.

The vergence is variable, and the folding axes are usually subhorizontal. There are several folding orders.

In Península Mitre, the clastic-eruptive complex of the FADC-ES is found (in Montes Negros) in apparent overlapping with the fossiliferous layers of the Late Cretaceous. This observation, which coincides with another one to the north of Lake Fagnano where the Tertiary rocks would be underlying the Late Cretaceous, allows the interpretation of the existence of a mega-fold that reversed the position of the large geological units. The presence of abundant fossils of approximately the same age toward the north (from the middle latitude of Montes Negros to Policarpo) and the south (Tekenika area) of the seemingly sterile layers of the central but supposedly older belt would reinforce the previous hypothesis.

(d) Cleavage. According to the attitude of the axial planes of the folds, a metamorphic structure has been developed. It is a slaty or axial plane cleavage, dipping in general toward the south, dominant to the point of printing geomorphological features defined in the landscape. The metamorphism is regional, of low degree (chlorite).

10.6.2 Minor Structures

The structural geology of the Deformed Complex of the Fuegian Andes is rich in microtectonic details. Accordingly, the minor folds and their relation to the major deformation structures have been mentioned and described in preceding chapters. Such is the case of the geometric arrangement in the critical location of the Río Pipo, whose kink-bands and folded quartz veins would be comparable to those of the Martial glacier area as well as the surroundings of Mount Olivia. The mention of deformative features in some minerals has also been mentioned, such as the folded titanite of the greenschists of Paso Garibaldi.

From the structural features mentioned, the real fact of the contrast between two fundamental characteristics arises: This is the folding of a large anticlinorium overturned to the north and large alignments between E-W and ESE directions that are parallel to the Fuegian cordillera. The folding is modern since it impacted layers of the Early Tertiary or, at least, the last folding is of this age.

Schematic geological profiles of the western and eastern sectors of the Deformed Complex of the Fuegian Andes are shown in Figs. 10.2 and 10.3.

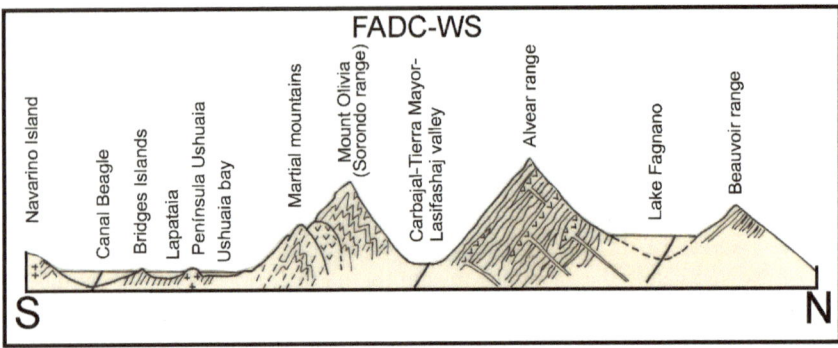

Fig. 10.2   Profil of the western sector of the Deformed Complex of the Fuegian Andes

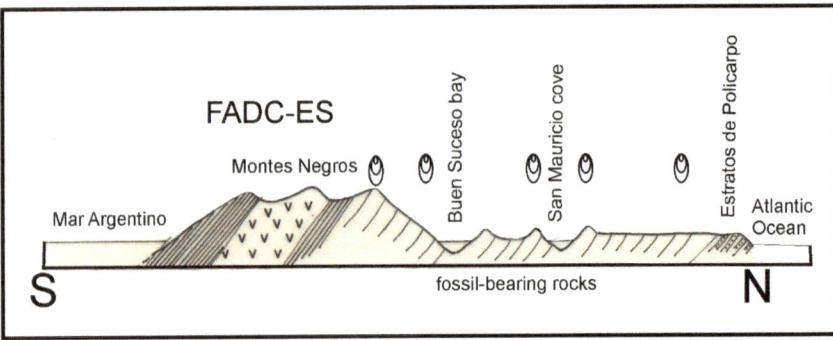

Fig. 10.3   Profil of the eastern sector of the Deformed Complex of the Fuegian Andes

References

Acevedo RD (1988) Estudios geológicos areales y petroestructurales en el Complejo Deformado de los Andes Fueguinos. Universidad Nacional de Buenos Aires. Biblioteca de la Facultad de Ciencias Exactas y Naturales. Ph. Thesis, 233p

González-Guillot M, Escayola M, Acevedo R, Pimentel M, Seraphim G, Proenza J, Schalamuk I (2009) The Plutón Diorítico Moat: Mildly Alkaline monzonitic magmatism in the Fuegian Andes of Argentina. J S Am Earth Sci 28(4):345–359

Martinioni DR, Linares E, Acevedo RD (1999) Significado de la edad isotópica de diques básicos intruidos en la Formación Beauvoir (Cretácico temprano), Tierra del Fuego, Argentina. Revista de la Asociación Geológica Argentina 54(1):88–91

Chapter 11
The Synthesis About the Formational Matter-Lithostratigraphic Units

The high nomenclatural complexity of the present analysis is shown by the fact that, before the Early Cretaceous, the geological literature records a lot of formational names, fruit of the contributions of many different authors. In all these cases, the creation of new names is justified both if they are being applied to lithologic groups or to formations, but as it has been said before, it makes the matter much more complex and also promotes the discussion of the identity of each unit. The last is clear enough and opens the chance, or perhaps the commitment of performing a simplification in agreement with the statements noted at the beginning of the present essay.

The complexity of the matter would make very difficult to perform a badly needed final simplification. This is of worth by itself, and the matter becomes wider when the identification of formations or units is doubtful in such a degree as to appeal to paleoenvironmental and paleosedimentary refinements in a restricted geographic environment. If the complexity of the overturned folds of many different orders is added to the above-mentioned facts, then the difficulties increase significantly, regarding the similarity of the rocks involved in the many different formations and units. Being much more appropriate, the denomination *Yahgan* to replace *Beauvoir*, *Serie Porfirítica* to replace *Lemaire Fm.*, leaving aside the concept of *Alvear Fm.*, and of *Esquistos de Lapataia* (Lapataia's schists) as a better denomination to other criteria or names, then such an explanation brings back to life the concept of the Fuegian–Patagonian Andes, now with very simple patterns of regional stratigraphic units.

This simplification is summarized in the following synthesis:

(a) The existence of a surface of discontinuity or unconformity that separates the underlying (b) Late Jurassic marine–volcanic deposits (mainly quartz porphyries) and the overlying (c) Early Cretaceous marine layers which can include the youngermost Jurassic (Tithonian–Necomian) period.

© The Author(s), under exclusive licence to Springer Nature Switzerland AG 2019
R. D. Acevedo, *Geological Records of the Fuegian Andes Deformed Complex Framed in a Patagonian Orogenic Belt Regional Context*, SpringerBriefs in Earth System Sciences, https://doi.org/10.1007/978-3-030-00166-7_11

Fig. 11.1 J. Benito walking across the Montes Negros (Mitre Peninsula). *Credit* R. D. Acevedo

   This great geologic entity does not lose validity neither due to local variations of stratigraphic differences, nor by the very important fact of keeping the Fuegian Andes marine system throughout the Late Cretaceous (Fig. 11.1) and the Early Tertiary.

   Table 11.1 is a combination of the stratigraphic chart and the geochronological scale. The thickness of the different bars is given neither by the lithic criteria, nor by the length of time involved. This chart is obviously not a great new improvement, but it remarks the need of using limited formational criteria like those used in this text. On the other hand, it must be acknowledged that in the Fuegian Andes there are more doubts than truly facts concerning the regional geology. Even the *Yahgan Fm.*, although usually well accepted, it is questionable because of the lack of chronological diagnosis in the original type area of the Argentine Andes. The post-Late Cenozoic glaciations (Bujalesky et al. 2008; Rabassa 2008) are deliberately excluded from this table for obvious time reasons.

Table 11.1 Geochronology, lithostratigraphic units, and geological processes in Tierra del Fuego

PALEOZOIC (Basement Zero)	JURASSIC l.s. (Basement One)	LATE JURASSIC-EARLY CRETACEOUS	LATE CRETACEOUS	EOTERTIARY	QUATERNARY/GLACIAL

Processes shown across the columns:

- MAIN RHYOLITIC VOLCANISM
- TECTONIC INSTABILITY (FLYSCH)
- OBSERVATION OF LEPTO-METAMORPHISM
- ANDEAN DIORITES
- LEUCOBASALTS-MELANDESITES-DIABASES
- ALKALI BASALT
- BASALTS
- FOLDING
- FRACTURING - OVERTHRUSTING

Yahgan Fm (Kranck 1932), Beauvoir Fm (Camacho 1948, Furque 1966), Serie Pizarreña (Harrington 1943), Monte Olivia Fm (Borello 1969), Estratos del Hito XIX (Doello Jurado 1926, Camacho 1949), Esquistos Antiguos (Bonarelli 1917), Clay-slate Fm (Darwin 1846) are included.

Serie Porfirítica (Harrington 1943), Lemaire Fm (Borello 1969, 1972), Alvear Fm (Borello 1969, 1972), Complejo Porfírico (Yrigoyen 1962), Lucio López Fm (Furque 1966) are included.

? Esquistos de Lapataia (Petersen 1949), Metamorfita Lapataia (Borello 1969, 1972), Lapataia Fm (Olivero et al. 1999) are included.
High Metamorphic Schists in Argentina (Kranck 1932), Paleozoic or Late Jurassic-Early Cretaceous. ?

References

Bujalesky G, Coronato A, Rabassa J, Acevedo RD (2008) El Canal Beagle: un ambiente marino esculpido por el hielo. In Sitios de interés geológico en la República Argentina. Tomo II: 849–864. Anales 46. Servicio Geológico Nacional. Buenos Aires

González Guillot M, Ghiglione M, Escayola M, Martins Pimentel M, Mortensen J, Acevedo RD (2018) Ushuaia pluton: magma diversification, emplacement and relation with regional tectonics in the southernmost Andes J S Am Earth Sci (in revision)

Rabassa J (2008) Late Cenozoic Glaciations in Patagonia and Tierra del Fuego. In: Rabassa J (ed) The Late Cenozoic of Patagonia and Tierra del Fuego. Elsevier, Developments on Quaternary Sciences, Amsterdam, pp 151–205

Chapter 12
A Simplified View About the Fuegian Andes and the Complex Folding of the Mesozoic and Early Tertiary Layers

The arch described by the Patagonian orogenic belt in Tierra del Fuego and the acute folding of the Mesozoic and earliest Tertiary strata are not two independent phenomena. The Patagonian Cordillera from Lake Fontana (44° 56′ S, 71° 30′W) to the south shows a simple architecture in strong contrast with the one of the Fuegian folding already outlined in the southernmost sector of the N-S belt, that is, north of the Fuegian arc. This matter has been discussed very few times in the geological literature; see, for instance, the theoretical–experimental work of Diraison et al. (2000) and references cited there.

The search for the causes that have produced these two structural features, arc and folding, has to take into consideration the undeniable fact that nature does not follow a strict one cause–one effect relationship as in physical or chemical experiments, but it is more a relation of groups of causes and groups of effects. Even the possibility of a convergence of multiple causes must be taken into account in this multiple game. The Fuegian arc and the building of a range with an E-W jointing orientation on one side the straight Patagonian Cordillera (N-S direction), and on the other side, it joins the great arc that encloses the Drake Passage and the Scotia Sea. It is a great geographic and geologic environment that surpasses without contradiction the hypothesis of H. J. Harrington (unpublished communication) of a subcrustal shifting movement as key factor in the origin of the Fuegian orogenic arc. Precisely, such an original hypothesis, in fact previous in time to the building of the plate tectonic theory, involved a division and a dispersion of the basement, a concept very reasonable enough.

The shift was in a non-perpendicular angle to the N-S Cordillera, i.e., SW or WSW in direction: An underthrusting or deep rigid movement forward worked in the orogenic warping, perhaps contributing then to the overturning of the Mesozoic strata. In such direction, approximately SW, it would be subperpendicular to the great alignment in almost a NW course noticeable in the Argentine rivers Pilcomayo (25° 21′ S, 57° 40′W), Bermejo (26° 51′ 47″S, 58° 22′ 58″W), de la Plata—Paraná (34° 30′ S, 58° 10′W), Dulce (28° 47′ 30″S, 63° 21′ 32″W), Colorado (36° 09′ 02″S, 70° 23′ 47″W), Negro (40° 48′ S, 63° 00′W), Deseado (47° 45′ 39″S, 65° 53′ 56″W),

© The Author(s), under exclusive licence to Springer Nature Switzerland AG 2019   107
R. D. Acevedo, *Geological Records of the Fuegian Andes Deformed Complex
Framed in a Patagonian Orogenic Belt Regional Context*, SpringerBriefs in Earth
System Sciences, https://doi.org/10.1007/978-3-030-00166-7_12

Fig. 12.1 View of Cordillera Darwin, Chile. *Credit* E. Paredes Morales

which would be tentatively interpreted as microtectonic plates of slow movement toward the SW.

The fracturing system is frequent in the subcontinental basement, the phenomenon not of uplifting of blocks, but of the not exclusive split of subhorizontal shifts. These spaces of basement like, e.g., those of La Pampa, areas of the basement north and south of the rivers Negro and Colorado depression, the isolation of Deseado's nesocraton, strongly suggest the possibility of detachments of "rigid islands" and that would be the case of the supposed deep one that could have advanced over the Patagonian Cordillera in an angle of effectivity as to produce the turn of the orogenic belt in initial development. They are in fact movements inside the South American plate, and it could be noted in the case of the uncertain migration of the *islas Malvinas* (Falklands) (Mitchell et al. 1986; Taylor and Shaw 1989; McKerrow et al. 1992; Marshall 1994; Richards et al. 1996; Stone et al. 2008). It would be like a neofracturing as in the former Gondwana, a late extension of minor shifts connected to the strong inertial condition of the continental masses. To this possible process of influence from the north, it is added the tectonic activity of the minor plate south of the Tierra del Fuego archipelago (Scotia Plate).

There are no clear elements to prove the above-mentioned movement, but this is not enough as to invalidate its existence as a logic hypothesis. And it would be also uncertain any attention to the metamorphites of the Darwin Cordillera (Fig. 12.1) as a shifted and finally uplifted block.

Concerning the acute folding of the Fuegian Andes, inclined toward the north semicircle, it indicates a linking with the building and development of the arc. The turn has been thus toward the foreland of the orogenic arc.

The folding is very well visible in the gorges crossed to the mountain structure. Open and irregular or disharmonic folds are added to isoclined folds. This repeated

Fig. 12.2 Acute Fuegian folding is independent of scale of observation (coin 27 mm diameter) *Credit* R. D. Acevedo

folding is the one that keeps away any precision concerning the original thickness of the layers. The metamorphism is of low grade, lower than that of the greenschist facies, with predominance of the deformation over the recrystallization and neomineralization. This speaks of a lower temperature in agreement with a reduced depth, which has meaning in the interpretations about the origin of the folding. As it has been said, the main characteristic is the overturning toward the north semicircle, that is, the foreland.

The folding is known to have happened in the Late Cretaceous–Early Tertiary times. The last comes from the evaluation of the layers involved. It was probably preceded by tectonic instability at the same time of the sedimentation in the marine basin.

In this way, there is the hypothesis of the slow movement advance of a crustal "island" toward the SW, involving the orogenic belt in its basin stage and finally in the rise stage when the subhorizontal pull was stopped by gravity.

The stratigraphic layers were uplifted giving origin to an inclination toward the northern semicircle. The acute folding, overturned (Fig. 12.2) and repeated, responded to a causal conjunction of this process and a possible pressure of the tectonic plate rim, as it was mentioned by Kranck in 1932, understanding that the folding responded to a tectonic action coming from the south.

Concerning the folding, the interpretation proposed here is that the final contribution of the gravity factor, i.e., in the *nappes* (obsolete term but pertinent here) of the Alpine tectonics, gave origin to a great overturned anticline, or more than one by *décollements* of the lithic mass.

This hypothesis deserves evaluation under the evidence coming from stratigraphic and structural research. Being it rejected, modified, or accepted, the result will be a contribution to the prevalescence of simplicity.

The sequence between the folding and the overturning is not unmistakable. It could be assumed that the folding started before the action of the gravitational cause, which emphasized the phenomenon, giving as a consequence a tightening of the folds. Such a process matches well with the low temperature suggested by the acting metamorphism related to such an intense folding which took place, ultimately, at a shallow depth.

The mentioned nappes were accompanied by *décollements*. Then it resulted in *nappes de charriage* with surfaces of movement turned toward the south semicircle by the design of the alignment of Lake Fagnano and possibly the Lashifashaj valley. Those same planes were finally the subject of horizontal shiftings, increasing and continuing the E-W lengthening of the Fuegian arc. It is evident the work of four forces or actions, respectively, in the following approximated ways: (1) S-N, (2) NE-SW, (3) W-E, and (4) vertical (Fig. 12.3). The first three had been the causes of the orogenic belt and the beginning of the folding, and the third force was responsible for the Quaternary tectonic activity of transcurrent movement (Lodolo et al. 2001). The fourth force is the gravitatory effect coming from the lithostatic pressure of the overturned mass. These positive forces were not simultaneous but they were superposed. The resultant from the gravity (action n° 4) worked by lithic weight after the overturning.

The frequently mentioned hypothesis of oceanic crust subduction as a tectonic factor finds its expression in the case of the Fuegian Andes in the subduction located south in the plutonic arc of the archipelago as a continuation of the belt of N-S subduction of the Nazca plate. That would involve the push to the north (force n° 1) as it was understood by Kranck in 1932. This has validity although no ophiolites from the bottom of the sea are yet known in the Argentine sector of the Fuegian Andes.

Another very different matter is the geologic meaning of the transcurrent fault and alignment of Lake Fagnano (the Magallanes–Fagnano fault) that was and is the subject of research because of its condition of being an active fault (Lodolo et al. 2001). But its importance in the understanding of the Scotia arc does not give place to linking with other geologic characteristics of the Isla Grande de Tierra del Fuego. It is the case of alkaline affinity of the Jeu-Jepén intrusion (Acevedo et al. 2000, 2004; Cerredo et al. 2000). In fact, the rock records only a later enrichment in potassium, frequent in the crystallization of many granites, a circumstance that by itself does not mean alkalinity. On the other hand, in these rocks there is absence of the needed feldspathoids that are a must to define, from the mineralogical evidence, an alkaline rock (Sørensen 1974), thus being logical the adscription of the plutonite rocks of Cerro Jeu-Jepén to a lateral exposure of the Andine batholite and not as a consequence of the activity in the Lake Fagnano transcurrent fault.

The metamorphism has also been linked with the Magallanes–Fagnano fault, but in fact, as it has been noted earlier in this work and also by other authors, the metamorphism is a phenomenon associated to the local folding of the area.

As a consequence, the Magallanes–Fagnano fault bases its importance by showing the lastingness of the stress produced by the third force.

In any way, the matter about the validity of the transcurrent fault of Magallanes–Fagnano is still open. If it is devoid of magmatic links, then it reduces, as consequence, the importance given in the present outlines about the definition of the

Fig. 12.3 Schematic illustration of the regional acting forces on Tierra del Fuego

Scotia plate. The above statement is in agreement with the simpler conception of the idea that the Magallanes–Fagnano fault is just an inner transcurrent fault of the South American plate.

The building process of the Fuegian arc has not had as a characteristic the volcanic activity, except in the case of those rhyolitic volcanics (quartz porphyries) of the named *"Serie Porfirítica"* or *Lemaire Formation*, which is connected with the Patagonian Jurassic volcanic activity and not with the Fuegian geologic characteristics in themselves. Then it is of a great value the identification of alkaline basalt (Acevedo 2016) located in a very limited area, inside the folded belt visible next to Mount Susana, near Ushuaia.

The plutonic axis, also curved, or in fact following the curve of the orogenic belt, is, as it is well known, located more to the south, accepting that the presence of such an important batholite has worked as a wall inhibiting the making of any symmetry in the folded group.

Even simpler descriptions about areal movements have already been outlined by Kranck (1934).

And from there to the present, theoretical interpretations of progressive deformation during growth of regional orogeny are put into consideration by Torres Carbonell et al. (2017). In the same way, the application of modern geophysical technologies that reflect the state of the art in our days reveals new aspects contributing to the understanding of the regional geology (González-Guillot et al. 2012; Bran et al. 2018).

References

Acevedo RD (2016) Alkali basalts and enclosed ultramafic xenoliths near Ushuaia, Tierra del Fuego, Argentina. SpringerPlus 5(1):1–5

Acevedo RD, Roig CE, Linares E, Ostera HA, Valín-Alberdi ML, Queiroga-Mafra JM (2000) La intrusión plutónica del Cerro Jeu-Jepén. Isla Grande de Tierra del Fuego, República Argentina. Cadernos do Laboratorio Xeolóxico de Laxe 25:357–359. A Coruña

Acevedo RD, Roig CE, Valín-Alberdi ML (2004) Lithologic types of Jeu-Jepén Diorite, Isla Grande de Tierra del Fuego. International Symposium on the Geology and Geophysics of the Southernmost Andes, the Scotia Arc and the Antarctic Peninsula. Bolletino di Geofisica teorica ed applicata. Instituto Nazionale di Oceanografia e di Geofisica Sperimentale 45(2):100–102

Bran DM, Tassone A, Menichetti M, Cerredo ME, Lozano JG, Lodolo E, Vilas JF (2018) Shallow architecture of Fuegian Andes lineaments based on Electrical Resistivity Tomography (ERT). Evidences of transverse extensional faulting in the central Beagle Channel area. Andean Geology 45(1):1–34

Cerredo ME, Tassone A, Coren F, Lodolo E, Lippai H (2000) Postorogenic, alkaline magmatism in the fueguian andes: the hewhoepen intrusive (Tierra del Fuego Island), IX Congreso Geológico Chileno, 2, Puerto Varas, Actas pp 192–196

Diraison M, Cobbold PR, Gapais D, Rossello EA, Le Corre C (2000) Cenozoic crustal thickening, wrenching and rifting in the foothills of the southernmost Andes. Tectonophysics 316:91–119

González-Guillot M, Prezzi C, Acevedo RD, Escayola M (2012) A comparative study of two rear-arc plutons and implications for the Fuegian Andes tectonic evolution: Mount Kranck Pluton and Jeu-Jepén Monzonite, Argentina. Journal of South American Earth Sciences. Amsterdam: Pergamon-Elsevier Science Ltd., 38:71–88

Kranck EH (1934) La Cordillera Alvear y el valle de Tierra Mayor. Revista Minera VI:49–64

Lodolo E, Tassone A, Menichetti M, Lippai H, Hormaechea JL, Ferrer C, Connon G (2001). The Magallanes-Fagnano Fault System in the Lago Fagnano, Tierra del Fuego Island: morphology and structure. EGS XXVI General Assembly. Continental rifts and passive continental margins Symposium Se.5.02. Niza

Marshall GEA (1994) The Falkland Islands: a key element in gondwana paleogeography. Tectonics 13:499-514

McKerrow WS, Scotese CR, Brasier MD (1992) Early Cambrian continental reconstructions. J Geol Soc, London 149:599-606

Mitchell C, Cox KG, Taylor GK, Shaw J (1986) Are the Falkland Islands a rotated microplate? Nature 319:131-134

Richards PC, Gatliff RW, Quinn MF, Williamson JP, Fannin NGT (1996) The geological evolution of the Falkland Islands continental shelf. in BC Storey, EC King and RA Livermore (eds) Weddell Sea Tectonics and Gondwana Break-up, Geological Society of London Special Publication 108:105-128

Sørensen H (1974) The alkaline rocks. Wiley, London, 622pp

Stone P, Richards PC, Kimbell GS, Esser RP, Reeves D (2008) Cretaceous dykes discovered in the Falkland Islands: implications for regional tectonics. J Geol Soc 165:1-4

Taylor GK, Shaw J (1989) The Falkland islands: new palaeomagnetic data as their origin as a displaced terrane from southern Africa. In JW Millhouse (ed) Deep structure and past kinematics of accreted terranes: Washington D.C., American Geophysical Union, Geophysical Monographs 50:59-72

Torres Carbonell P, Cao S, Dimieri L (2017) Spatial and temporal characterization of progressive deformation during orogenic growth: example from the Fuegian Andes, southern Argentina. J Struct Geol 99:1–19